KT-210-483

1 x 3 shear blackface tups,
Connachan, by Crieff, Perthshire,
1st December 1998

RURAL PORTRAITS

Many of the breeds described in this book are deemed to be rare or endangered. As farming practices have changed, some of the older native breeds have been supplanted by faster growing types from Continental Europe or North America to suit the changing market needs of today. Yet these old breeds have qualities that we cannot afford to lose – adaptability to local habitat, easy-care, low input, disease resistance, special wool qualities or other distinctive factors such as high eating quality and extra taste.

The charity, the *Rare Breeds Survival Trust* (*RBST*), is dedicated to conserving all rare breeds of farm livestock from these isles. They currently have over 70 breeds on their Priority List of cattle, sheep, pigs, goats, horses, ponies and poultry. Such breeds have dedicated enthusiasts who keep them going, some of whom are described in the text. Rare breeds are as important a part of our heritage as our ancient buildings and art collections. If you would like to find out more about them or how you could help by joining the RBST, visit their website www.rbst.org.uk or ring them on 024 7669 6551.

RURAL PORTRAITS

Scottish Native Farm Animals, Characters and Landscapes

POLLY PULLAR Polly Pullar

Illustrated by
KEITH BROCKIE Keith Brockie

The author's Shetland sheep

Langford Press

Langford Press
Book Corner,
2, High Street, Wigtown,
Wigtownshire, DG8 9HQ

Email sales@langford-press.co.uk

Graphic design and layout by Lesley Beaney
Typeset in Minion and Cerigo by
Pioneer Associates, Camserney, Perthshire
printed and bound in Singapore
under the supervision of
MRM Graphics Ltd, Winslow, Bucks, UK.

A CIP Record for this book is available
from the British Library

ISBN 1-904078-06-0

Contents

Hebridean ewes and lamb *(see p. 89)*

For my wonderful son Freddy,
Mum, Mike and Susie,

With all my love

And for James Biggar
and
Sandra Nicholls

for whom I have great admiration

Bowhill, Selkirk, Scotland, TD7 5ET

This is a book of surprises – all nice ones! Nothing like it has ever been done before. It is a unique record of Scotland's native breeds of farm livestock and much more. It vividly portrays with pen and paint, choice views of rural life throughout the whole country, including the native breeds, associated landscapes, wildlife and colourful human characters.

It is a book for all ages from 5 to 105. Each chapter is a separate delightful story and can be read in any order. The brilliant illustrations by Keith Brockie provide the first striking impression. With outstanding talent he presents people and animals in their natural surroundings in apparently three dimensional form and detail that no camera could surpass. One can see into the eye of a bull and can almost feel its breath and smell its aroma. But this is only half the story.

Polly Pullar's verbal portraits of our native creatures in their true setting, total strangers to so many, bring them vibrantly to life. The charm and humour of her writing can hardly fail to bring a smile of pleasure to the most dour worshipper of pavement life in town or city. This is an important and accessible book which invites the reader into a wonderful world, not of make believe, but of reality. It deserves to find its way into every home and certainly every school and public library.

The timing is spot-on, just when the survival of some of our native breeds is under threat and when agriculture is undergoing tremendous change. These are the animals that over centuries have become acclimatised to our very diverse parts of Scotland. They are unique. Once they go, the basic breeding stocks can never be replicated.

Today only 5 per cent of our population have their roots in the countryside, so it is understandable that so few of the other 95 per cent have more than a fleeting interest or understanding of what happens there. Sadly the gulf in understanding seems to be widening. Although there is a widespread agreement that our landscape is about the most beautiful to be found anywhere with huge variations within a short distance of each other, it is generally assumed that it will automatically remain just as it is, frozen in time. But although the Almighty gave us the hills and the glens, it is man who has clothed them with their conspicuous adornments, the ever changing cropping and trees, single, in hedgerows, clumps, shelter-belts, small woodlands and great forests. Even apparently bare hills change greatly according to grazing regimes, sporting interests, bracken control and man-made intrusions like electricity pylons, telephone lines, mobile radio transmitters and wind farms.

Almost all changes, good or bad, are in the hands of man. Unless he manages the land in a responsible far-sighted way, they will be for the worse. If therefore, decisions are dominated by the 95 per cent urban minded population, as they generally are, without consulting the countryside 5 per cent, much of what is so often taken for granted could be swept away for ever.

Country people are feeling the strain following Foot and Mouth problems and a sense of isolation. As President of the Royal Scottish Agricultural Benevolent Institution, I express deep gratitude to both Author and Illustrator for offering a portion of the proceeds of this book for the essential support given to rural workers of every kind in their hour of need. The Foot and Mouth disaster seriously affected so many who may require help for years to come. The Institute was immensely grateful for the colossal response to our fund-raising appeals in 2001 and now I express our sincere thanks for this generous offer to help us to carry on in the longer term.

The need to build bridges in understanding between town and country has never been greater. This is where Rural Portraits can play such a very significant part. It could bring a new light to those who live in an urban twilight. It will be illuminating for every age group, and should be made widely available.

Buccleuch.

The Duke of Buccleuch, KT

Acknowledgements

COUNTLESS people have helped and encouraged us with this book. All who are included have generously given up their precious time. We have sat round their kitchen tables, partaken of their delicious soup and hospitality while I in turn plied them with hundreds of questions. The Scots must surely be the world's best soup makers. On top of this intrusion they have willingly moved animals into perfect settings, and had them ready in fanks and outbuildings waiting to be sketched or photographed. Special thanks to Brian Ridland of Shetland, who on a tempestuous day, extricated himself from work and placed their beautiful flock of katmogit Shetland sheep on a rock surrounded by wild sea to create a stunning image.

We have been constantly guided by Lesley Beaney and Jan Dunn, of Pioneer Associates, Camserney. They deserve very special thanks, as without their assistance, advice, help and patience we would have been lost. Pippa Catling has corrected my bad punctuation and dealt with errant commas and shimmering sunsets uncomplainingly. She has given me hours of her time, and made endless helpful suggestions while Tim Wynne-Williams made copious cups of tea and coffee. David Leggat at United Auctions has given us great moral support throughout, despite being one of the busiest of people. I also thank Joe Harris whose advice bore great fruit. Thanks are also due to the Breed Societies, and The Rare Breeds Survival Trust who took the trouble to check the scripts.

I would like to thank my stepfather, Mike Thomson, and finally and most importantly, my mother, Anne, who has instilled in me the love and the passion that forms the background to this book.

Keith would like to thank Margaret Calder and his parents, Derek and Anne.

The views expressed by the people in this book are not necessarily those of the author.

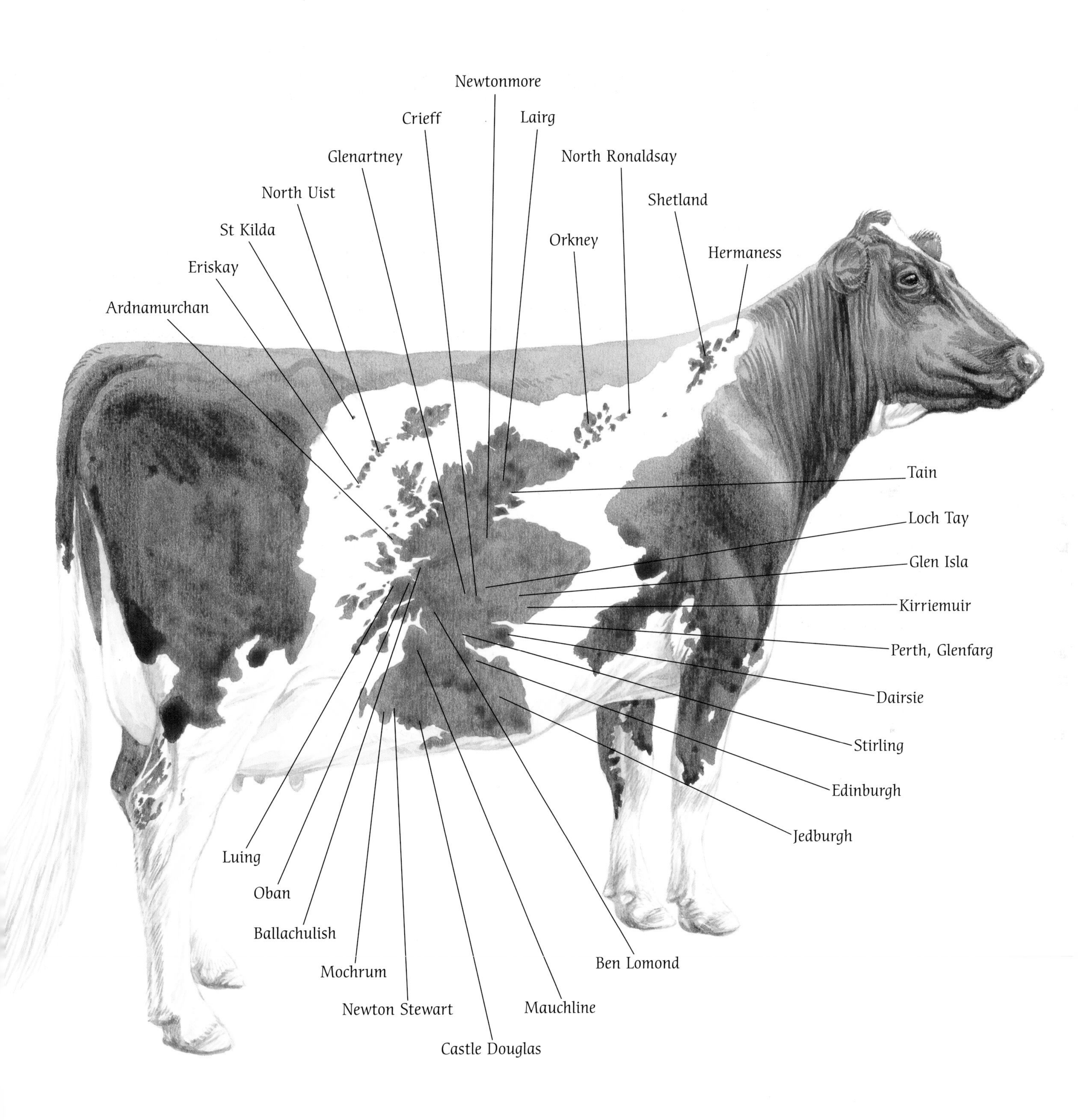

Map showing some of the places visited while compiling Rural Portraits

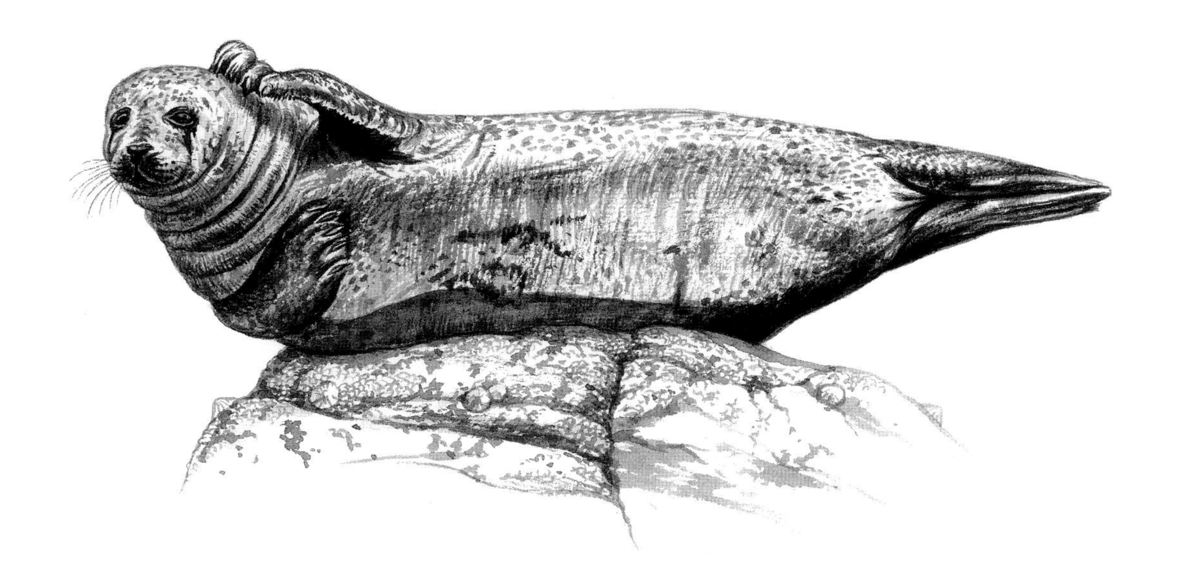

CHAPTER ONE

Ardnamurchan – Early Days

ONCE you have travelled on the Corran Ferry that ploughs its way back and forth through the racing sea currents of Loch Linnhe, the drive down the Ardnamurchan peninsula to the most westerly point on mainland Britain is quite breathtaking. However, a well-known firm of furniture removers found it very daunting in the mid-1960s as they battled to negotiate its vertiginous rocky corners. The road wends its way through woods of stunted oak, hazel and birch embellished with greeny-grey lichen and moss. There are sheer drops into Loch Sunart, where seals haul up on the barnacle-encrusted islets. Deep peat bogs on either side cushion the craggy ledges where clear water cascades into little burns that run like veins throughout this area of unusually high rainfall.

On their arrival many hours later, the fraught lorry driver and his crew fell into our newly acquired hotel bar in Kilchoan. Buffeted by a tempestuous storm and evading disaster in a deep peat hag, they were left with a lorry that was in sore need of a re-spray and were clearly in no hurry to return. Jangled nerves had left them with an insatiable thirst. As my mother said, 'It took them nearly a week to recover from both the journey and their excessive thirst.' From their base in urban England, little could have prepared them for the hazards of the single-track road to Kilchoan, where dozing sheep lie cudding and immovable on the pitted tarmac and herds of deer leap out of the gloaming into the path of passing vehicles. Late in the evening, in the public bar, a broad Liverpudlian accent was overheard to ask, 'What on earth made them want to move to a God-forsaken place like this in the first place?'

This, the home of the golden eagle, buzzard, pine marten, fox, otter and badger, was to form the basis of my greatest love of all. It was to shape my thinking and fire me with a burning passion for wildlife, the wilderness and the way of life in rural and agricultural communities.

The ceilidh that developed in the public bar the night our furniture arrived in Kilchoan, as the

(above) Common seal

removal men soothed their frayed emotions in the wake of their hair-raising journey, was the first of many. As our new lives in Ardnamurchan unfolded and we became involved in village life, I grew to love my conversations with the colourful characters and their tales that inspired me and gave a wonderful insight into their lives and livestock. Sadly, the oral tradition is dying very quickly.

When we moved to Ardnamurchan there was no mains electricity. The crofters and villagers had tilly lamps and gas lights, while our hotel was run on a very unreliable generator which often broke down, plunging our visitors into darkness. Lack of work in remote places has led to the departure of many of the young people from such far-flung corners as Ardnamurchan. There are fewer children in the little schools and more importance is obviously put on the National Curriculum than on the natural world. Today, rural children are thought to need special facilities such as swimming pools, sports halls and computer rooms. We never had these things, yet when the coal puffer came to the village pier at Kilchoan twice a year, we were given a day off school and rode on the tractors and trailers as they went round the village delivering the coal. We competed to see who could become the

Mingary Castle

blackest. This kind of fun would be totally unheard of now. Our growing expectations of life have altered our attitudes, and many of the benefits of living in a small rural community have been lost.

I had quite the best childhood imaginable. Safe from the hazards faced by street children, I sought my entertainment knee-deep in the rich rock pools round Kilchoan's varied shoreline, or fishing and guddling in the brown hill burns. We collected treasures, such as prickly purple and green sea urchins, brittlestars with waving tentacles, goosefoot starfish and oozing jars of frog spawn. Days were spent with the shepherds at the sheep fank, jostling with the Blackface sheep that have remained one of my favourite breeds. Time revolved round lambings, calvings, clippings and dippings. There were rescues too, when cattle were bogged in a peat hag or terriers had gone to ground in a fox's cairn. I loved going with the stalkers, walking miles across the open hill on Ben Hiant with a stick and a piece bag, often being soaked to the skin. Dozing in the autumnal sunshine amid the roars of the rutting stags, with the Sound of Mull shimmering far below, was awe-inspiring.

The return to boarding school was always quite horrible. Gone was my freedom. In its place nothing but soulless lessons with dull teachers, who simply did not comprehend where my heart lay. How could school compare to days spent on the foreshore collecting drift wood and fascinating remains washed up by the tides, or hiding in the bracken watching the wild red deer throughout the seasons? At home ravens, buzzards and eagles were constant companions. Looking out of my bedroom window in the morning, my first view was of the ruin of Mingary Castle and the glorious Sound of Mull in all its moods, while otters were frequently seen on the shore below.

Ardnamurchan is often shrouded in low mist and drenched in impenetrable damp for weeks on end. Some visitors can be unlucky and see no further than the bonnet of their cars. The islands of Rum, Eigg, Muck, Skye, and Mull can be obscured totally in the intense grey and horizontal rain. Everything has to be well battened down as vicious gales sweep in off the Atlantic, smashing caravans to smithereens and removing roofs with an astounding ease. When my mother bought herself a new greenhouse it quickly became a victim of the gales of the autumnal equinox, and pieces of it were last seen heading in the direction of the Isle of Coll. Some days the winds are so violent that even stepping foot outside the door becomes hazardous. Then, in summer, black clouds of the most voracious midges on earth can make damp mornings and evenings a living hell.

Nevertheless, Ardnamurchan boasts some of the loveliest scenery in the world, and when the weather is obliging it is a wildlife lover's paradise. The flora and fauna are unequalled. There are heavenly shell-strewn beaches where flocks of oyster-catchers and waders play in the tides and lithe otters waltz in the bladder wrack. The call of the curlew hangs hauntingly over the lonely ruined crofts, where once whole families lived. By June, when the red deer are calving, the hillsides are carpeted in a dense mat of wild flowers, and above the tide-line swathes of flag iris burst into an explosion of lemon. In the craggy gullies, inaccessible to grazing deer and sheep, a wealth of plants, trees, ferns, and bushes grows in abundance, providing little havens, like magical undiscovered gardens. Orchids grow in profusion and the stunted wind-battered trees become gnarled and bent into a myriad of extraordinarily beautiful shapes. Hill lochans covered with bobbing water lilies and edged with reedy grasses echo to the eerie calls of the nesting red and black-throated Divers.

Running a hotel proved to be totally exhausting, but provided a constant supply of hilarious incidents, inspired by the local characters and the visitors. The road from Ardgour to Kilchoan has since been totally upgraded, though it is still narrow in places and it is useful to be adept at reversing. However, in the 1960s, it was negotiated as little as possible.

Finding fresh fruit and vegetables for the hotel was quite a task. Glasgow Fruit Market might as well have been as far away as the moon. Numerous lorry drivers were floored by the switchbacks and hairpin bends. Many ended up in the peat hags, while others burst tyres on the protruding rocks. By the time any fresh produce had survived the journey, it looked as limp and unappetising as the poor drivers. Some provisions were sent by ferry from Oban, first stopping in Tobermory on the Isle of Mull, where they were often waylaid. However, the daily bus left the village each morning, returning with milk, bread and mail. 'Hughie the Bus' deserved a knighthood by the time he had withstood the road for a quarter of a century, and indeed received the BEM for his endeavours.

Our cook, Betty, was the key to the smooth running of the hotel. Without her incredible ability to concoct something out of nothing, my parents would not have survived life as hoteliers. Soon after we moved to Kilchoan, our first pet lamb appeared. Lulu was a sickly Blackface lamb belonging to a neighbouring crofter and, like so many of her kind, survived despite the odds. She soon became heavily imprinted on us all and a much-loved member of the family, featuring in a newspaper with a photograph of her following my mother up the hotel stairs with the guests' early morning tea. As Lulu grew up, she became enemy number one to Betty.

The ovine forays into the hotel, which had once proved so popular with visitors and press alike, soon became dreaded by Betty. First there was the tray of delicious newly baked shortbread. It lay golden and sugary, awaiting the afternoon tea rush one wet summer's day. Lulu demolished every neatly scored square, and stuck her sugar-coated muzzle round the kitchen door as she departed, leaving nothing but a trail of droppings, like black currants, in her wake.

Ominous little black currants often blazed a trail to the bedrooms too. Weary guests, on being shown to their room, found Lulu and her lamb happily chewing the cud with their heads against the comfortable candlewick bedspreads. However, it was the incident with the peaches that caused the

Ardnamurchan Point

most friction. A whole tray of perfect blushing peaches had arrived safely from Fort William, and peach melba was on the dinner menu. Triumphantly, Betty went into the food store in the late afternoon, to find that all that remained of this dreamt of treasure was a stone neatly spat back into each shaped cardboard section where once there had lain a prize. Despite her diminutive size, Betty's wrath was gigantic, and Lulu was chased up the road with a large carving knife accompanied by a gale of expletives.

The District nurse also had a pet lamb, by the name of Paddy. He and Lulu joined forces and locked themselves in the gents' outside lavatory. Once the terrible mess had been cleaned up, some bright spark inscribed 'Paddy loves Lulu' in big letters up the wall.

Numerous other animals punctuated our time in Ardnamurchan. These creatures made my departures back to school even more loathsome. There was Winifred, a guinea-pig that I acquired on a visit to England. Returning home as an 'Unaccompanied Minor', with my precious cargo in a box, my father, now separated from my mother, had been forced to fill in endless forms that had to accompany a jet-setting guinea pig. Flo was a naughty spaniel who was not averse to stealing sirloin steaks from behind Betty's back. She once raided the neighbour's hen run, leaving several of the occupants 'oven-ready'. There were Ginger my pony, and Seamus, a yellow dun Highland pony, who

successfully managed to sink up to his belly in a bog with my stepfather on board. Inevitably, this was the only time we had ever enticed him on to a horse. We also had a Highland heifer calf called Flora, and many others.

The farm animals of Ardnamurchan became of constant interest to me. There was a pedigree fold of Highland cattle, and a herd of hardy, mossy Galloways that thrived on the poor grazing and withstood the abuse the climate constantly hurled at them. I loved going up to the bull-pens to watch the steaming breath of some of the huge occupants as they ate their hay. Compared to the vast continental creatures that are at the forefront of agriculture today, those bulls were tiny, though at the time they seemed enormous.

I remember taking a short cut to the shore through the middle of a field of bulls with three older boys. Suddenly, a rather fiery Aberdeen-Angus began to paw at the ground, bellowing in a threatening manner. Within seconds all five bulls were hurtling towards us. I ran as fast as my little legs would carry

Rum and Eigg from Achateny

me. but saw to my horror that the boys were miles ahead. With a thundering of hooves and cascades of mud the bulls shot past me. Breathless, I realised that I was now running behind them. I never saw three boys leap over a barbed wire fence so fast. One of them left the seat of his breeks hanging like a flag. After we were all reunited, we rolled on the ground and giggled with relief.

Though we were largely cut off from the outside world, village life was never dull. People came and went. 'Vic Pictures' brought wobbly black and white films to the village hall several times a year. Once the films were finished, another ceilidh might develop. As we escaped outside for a breath of fresh air in between dances, the rasping calls of the corncrake drifted up from the beds of marsh marigolds close by. Travelling salesmen frequently came to the village. 'Peter the Darky' brought a large selection of totally un-waterproof rainwear and other clothing in various unsuitable shades of pastels, much to the delight of the frustrated shoppers.

One of the best-loved characters was Lizzie, who kept a tea-room and helped in the hotel. Her flock of Blackface sheep wandered round the village munching any colourful blooms that showed their faces. She took all the scraps from the hotel for her beloved hens, the eggs of which had quite the deepest orange yolks imaginable. It must have been their seaweed diet, for they were often scratching on the shore. Some could even be seen peeping from the windows of her little house. How I adored to visit Lizzie! The tea-room was one of the favourite haunts of the visitors too. All the delicious home-baking was priced at 1/6d, for the defunct till she had acquired from the hotel had jammed at this price.

Christmas in Ardnamurchan was still little celebrated in the 1960s. My mother remembers looking out on our first Christmas morning, and seeing the hearse filing down to the pier to catch the Mull ferry. Funerals even took place on Christmas Day. New Year was a different matter, and the celebrations continued for weeks on end. Hogmanay was spent at the lighthouse at Ardnamurchan Point. Here everyone gathered and sang Gaelic songs over the radio to all the other lighthouse keepers on islands such as Barra, the Uists, Lewis and Tiree. Often the weather added to the atmosphere, as massive breakers smashed against the rugged grey mass of Britain's most westerly point. My mother filled the 12-seater Land Rover with those who were game enough, and ventured to the lighthouse once the bar was shut. The party returned at dawn in an alcoholic haze, hungry for bacon and eggs, more singing and eternal merry-making.

Boat trips too were a feature. On glorious summer evenings we fished for writhing silvery mackerel that we grilled with butter and oatmeal. I was so excited on seeing my first killer whale, while we were trolling for mackerel, that I nearly fell out of the boat. Sometimes dolphins and porpoises followed curiously in our wake, treating us to fine displays of their lissom acrobatics.

One memorable adventure involved a persistent fox that had been creating mayhem in the hen house. It had defied all attempts on its life, until one morning my mother returned home for breakfast and triumphantly announced that she had shot it. It was mortally wounded but had crawled into an inaccessible gorse bush where, she said, it had expired. Shooting a fox in Ardnamurchan was cause for celebration since they were regarded as vermin, and a bounty was paid for their brushes. Later in the day I found my parents in a very buoyant mood. They had cracked open a bottle of whisky, and were dreaming up plans to hire a six-seater plane to visit a long-lost friend who had escaped to live on a tiny island off Benbecula. Given the considerable logistical problems, a holiday tour operator would have been most impressed by the way my stepfather overcame all the immeasurable obstacles that made reaching Benbecula so difficult.

We set off on the daily passenger ferry, the 'Loch Nell', to Mull. A taxi took us from Tobermory to the tiny airstrip at Glen Forsa, where we met our very dapper pilot, complete with epaulettes and captain's cap. The first hazard was a flock of sheep that were enjoying the sweet, neat grass on the airfield, while the second was a group of islanders intent on us joining them for a party in the 'terminal', a wooden hut. Finally, the runway was clear, and we were airborne. My mother hates flying at the best of times, and her enthusiasm for the trip dwindled rapidly as we bumped our way up through the air pockets above the glorious mountains and glistening sea far below. It was dramatic. White shining beaches emerged through gaps in the wispy cloud, and green patterns of fields and crofts sprawled over the patchwork landscape. We circled furiously at Benbecula as our overwrought pilot awaited permission to land, finally coming to earth with a bounce.

Colin, my stepfather's friend, was waiting for us with his ancient decaying Land Rover. My step-brother James and I clambered into the back where a squirming collie bitch licked my face, her whole body sinuously wagging in greeting. There were coils of tarry rope, a spare tyre and a huge parcel containing an enormous leg of lamb. 'Thank heavens that dog hasn't devoured our dinner,' laughed Colin as we squeezed ourselves in with all the tackle.

After a few miles cramped in the Land Rover, Colin sped down a muddy track to the shore where we fell into a small but solid-looking boat, held together with bitumen. Sticky tar oozed from her

Red deer calf

battered sides. My poor mother was looking unenthusiastic again. The icy green water was choppy when Colin pushed us out in the gear-laden boat and exclaimed heartily, 'Damn, I knew I forgot something; these boots leak like a sieve.'

There was a house at either end of the little island and at one of them Colin's elderly mother was awaiting us in their doorway, with a small pet lamb at her heel. Soon she had put the leg of lamb safely in the range, and Colin was out digging potatoes from his lazy bed. Meanwhile Knickers, the pet lamb, was running round the living room leaving pools while the collie watched her every move. 'We called her Knickers because it would have been better if she'd had some,' laughed Colin's mother, who spent most of the evening mopping up puddles and searching for her elusive spectacles.

Meanwhile, Colin and my family were catching up on years of tales while whisky was flowing as freely as water from a Highland burn. Dusk fell. A wisp of peat smoke drifted up from the little chimney, and the last flocks of herring gulls flew home to roost. A thin moon rose in an opaque sky as the waves lapped on the shore below the croft. I sat on the step with my two animal companions, listening to the huge silence.

By the time we remembered the leg of lamb in the oven it was a shadow of its former glory, but the tatties were succulent and earthy. 'Have plenty of butter with them,' Colin insisted as he ladled butter and vegetables on to my cracked earthenware plate. After supper a dusty squeeze-box emerged from a spidery cupboard beneath the stairs. The night was filled with sounds of a diverse orchestra that lacked musical talent. At midnight I went up the narrow staircase in search of my sleeping bag. The stars twinkled clearly down from the skylight window and I fell into a deep sleep.

As daybreak unfolded its grey blanket over the croft, I was rudely awoken by rain showering down through a hole in the glass. No wonder the stars had seemed so bright. A steady rhythmical snoring drifted up from below. The party had ended in mid-flow; James lay on the sofa sound asleep, while

Camus nan Geall

Colin sat at the table, his squeeze-box still open and groaning gently in time to his snores. My step-father was still clutching his empty glass. The scene was peaceful. I pulled the door to silently, and went out to take the delicious salty air.

I met my mother on the headland. A small byre housed the 'thunder box'. This 'loo with a view', though far from comfortable and lacking all the luxuries of twenty-first century plumbing, was uniquely spectacular. We returned from our ablutions to awaken the men from their peaceful slumbers. Colin already had the kettle brewing but had laced the teapot with a liberal splash of whisky. This revolting mixture received a lukewarm response, and a virgin brew was also made. After breakfast a shriek of horror announced that we had forgotten our smart pilot, ticking away his time at Benbecula at vast expense. Clothes were hastily thrown into bags as we dashed for the boat. She lay like a vast stranded whale far above the tide-line. Though we pushed and we shoved, she was immovable. My wellies filled with cold water while the sea remained an unreachable goal and, with much gnashing of teeth we were forced to admit defeat.

The only telephone was at the other end of the island. We strode forth like intrepid explorers across the spongy landscape. A familiar red phone box stood incongruously next to the neighbour's house, and we frantically rang the airport to explain our predicament. Meanwhile the occupants of the house were brewing up and lured us in for tea and further drams. Their hospitality was remarkable, but the return route to Colin's croft seemed more circuitous as we merrily wound our way round the peat bogs. The boat, meanwhile, was now floating innocently. With more than a tinge of sadness, we chugged away from the tiny island, waved off by Colin's mother and the lamb.

Our pilot was clearly unamused and was very icy with us. However, he thawed a little on the return flight and flew us over Ardnamurchan so we could see our own house. Our sheep and cattle looked like model animals from a children's farmyard. The pilot's temper was refuelled at Glen Forsa as the flock of sheep refused to remove themselves and we were again forced to circle. A county council lorry too was parked in the middle of the runway, its occupants involved in a tea-break elsewhere. Soon the ground staff, an old lady waving a red and white gingham tea towel, emerged, and the sheep dispersed like white cotton wool balls. After a lengthy conversation, two men loitered across the strip and drove the lorry away in a cloud of diesel fumes. By now our pilot was as cross as a bag of weasels, 'The Authorities will be hearing about this.' He had definitely lost his sense of humour.

Back at home, Benbecula seemed a lifetime away and we were all ready for a proper night's sleep. However, next day, as dawn broke pinkly over the farmyard, my mother's 'dead' fox limped stealthily out of the hen house, its face adorned in feathers. The celebration of its demise, had after all been the reason for our excursion to Benbecula.

Ardnamurchan still moves me more than any other place. There is something very reassuring about being able to return to find all the magical places of my childhood intact. Keith and I returned one February. At Achateny Beach as shafts of brilliant light danced all around us, banks of hail drove in off the sea, and the winter clad peaks on the distant Isle of Rum teased us with their shyness, we had one of the best otter encounters of our lives.

We first saw the two heads from a considerable distance as a mother otter and her three-quarter grown cub dived and splashed in the bay. Two great northern divers came and went on the waves as the heads and tails of the otters emerged at intervals. We hurriedly scrambled across the lethally slippery rocks to have a better view. The wind was in our favour as we crouched breathless behind a barnacle-covered rock, urging our excited dogs not to move a muscle. We waited silently. The female appeared only yards from us so that her whole head filled the frame of the binoculars. Beads of salt-sea droplets clung to her whiskers as she swam through the effervescent water with a shining brown lump-sucker in her half open mouth. We heard the sharp crunching of her teeth on the rubbery flesh, and the wet splash as she crawled up on to the rock before us. Then she was gone again in a mist of spray, re-emerging on the far side, bobbing and floating in a small tidal inlet and raising herself out of the boiling water to have a better look at us. She snorted a warning, then vanished.

From the rocks on the other side we heard the peeping alarm call of the juvenile otter and saw it ambling over the forests of seaweed until it curled itself up into a semi-circle, dozing cautiously. The agitated call continued as the mother swam elusively under the water. I clutched our quaking whimpering dogs as Keith crawled as close as he could to the oblivious cub. I watched both of them just a few yards from me, as he slithered across the wet rocks to within a couple of feet. Even when it saw him, their meeting seemed to be suspended in slow motion for several minutes, and neither moved. Then with a flurry of shining wet pelt, the cub dashed round the rock only to reappear a foot away to study Keith inquisitively before hurriedly sliding back into the sea.

Later in the afternoon, we watched another adult otter, large, round, and curvaceous, ambling along the sand, where rivulets of fresh water from the burn had etched out their pattern of ridges. Eigg and Rum peeped out from the chasing clouds as the otter disappeared, leaving nothing but its tracks on the seeping wet sand. Soon the tide would return to erase them. Memories of childhood sightings in this special place came flooding back through the curtains of reflected light playing on the water, highlighted by fleeting rainbows. Almost too much for the human soul to bear.

great northern divers with
otters in the distance
swimming in with a
lumpsucker fish
mother otter has caught our scent
juvenile asleep amongst seaweed
Achateny, Ardnamurchan
25th February 99

CHAPTER TWO

Heather Hiking with Galloways

I AM sitting in the new farmhouse at the Jaw Farm, Fintry. From the kitchen windows the distinctive shape of Ben Lomond lies snow-capped in the distance, while beneath the farm's lower fields, Fintry Church is the nearest neighbour. On a huge television screen a video of an old wobbling black and white cine film is running, and on it a group of cheery children cavort round an idyllic farmyard, surrounded by dogs, fowls, sheep, goats, and ponies. One particular child stands out. With a grin as broad as a cart-horse's back and a cheeky mischievous air, he is playing to the camera, and even at this youthful stage of his life has developed a real farmer's gait. Now he is appearing again, a few years later, as a teenager, clutching a large crook and wearing a huge pair of tacketty boots. A collie is working a group of ewes that run before him. John Maxwell clearly mastered the *heather hike* when he was very young. This is the nickname given to a long rocking stride useful for covering tough heather tussocks, frequently adopted by shepherds and keepers wearing traditional boots with large turned-up soles and toes. The broad grin and air of fun that surrounds the child in the film has little altered in the man today.

John Maxwell claims that he is uneducated. If by this he means he has few formal qualifications, it has certainly never held him back, for there are no flies on John. I have always admired him not only for his wit and charisma, but also because he is astute and clever, and is without doubt one of the hardest workers I have come across. In Maggie, his wife, there are the very same traits.

John's main passions are Blackface sheep and Galloway cattle, and he is famed for his expertise with both breeds. He was born near Thornhill, Dumfries, where his family had been tenants of the Buccleuch Estate since 1747. On a hill farm with 60 arable acres, and 760 Blackface ewes, he began his farming career working for his father, and later certifying animals for subsidies. He met Maggie at the

(above) John Maxwell with Jaw Hiawatha

Jaw Hiawatha

LIMITED EDITION PRINTS & CARDS
from the book
RURAL PORTRAITS

A

C

B

D

F

E

G

I

H

LIMITED EDITION PRINTS
PUBLISHED BY THE FEARNAN GALLERY

A set of limited edition prints of Keith Brockie's paintings from the book, **RURAL PORTRAITS**, is available.
Prints **A, B, C, E, F, G & I** are reproduced on the highest quality matt 300gsm neutral pH pap to meet the requirements of the Fine Art Trade Guild and to ensure permanence in limited editio of 850, signed and numbered, printed by the *waterless* lithographic process.
Prints **D & H** are printed by the Giclée* method, an 8–colour process on archival fine art pap using light fast inks. The superb quality image has all the tonalities and hues of the original. Ea is limited to 90 signed and numbered prints.
For further reference the relevant book page is given in brackets. The dimensions indicate in heig x width the actual image size plus a generous white border.
Postage/packing: Prints will be sent flat packed: **A,B -** £6 each, **C, E, F, & I -** £8, **D,H** - £10.

REF	TITLE	PAGE REF	SIZE	PRICE
A	Peacock butterfly	Page 22	20 x 15	£25
B	Small tortoiseshell butterfly	Page 23	20 x 15	£25
C	Blackface tups	Endpapers	24 x 37	£40
D	North Country Cheviot ewes *	Page 150	42 x 31	£110
E	Highland cow, North Uist	Page 82 / cover	32 x 26	£40
F	Scots Dumpy hen	Page 86	45 x 31	£40
G	Scots Grey cockerel	Page 87	45 x 31	£40
H	Beef Shorthorn bull *	Page 46	50 x 34	£120
I	Blackface ewes	Page 38	35 x 22	£40

Deluxe gloss colour greeting cards size: 20 x 15 cms are available printed on 380 gsm card featuring images **A, B, E, F, & G (£1.60 each)**
All 6 images - **£8 plus £2 p& p**
Visa, Mastercard, JCB, Switch, & Delta credit cards accepted.

Contact: FEARNAN GALLERY, by ABERFELDY, PERTHSHIRE, PH15 2PG
Tel/Fax: 01887 830609 www.keithbrockie.co.uk

Young Farmer's Club, and quickly realised that he should snap her up straight away. They are very well matched.

It was in the market in Castle Douglas that he bought his first dun Galloway heifer, little knowing at that time that he would remain faithful to the dun-coloured Galloway, eventually building up an impressive herd. The Galloway was very much a part of the surrounding countryside, being perfectly suited to the wetter climate of the west, and requiring no intensive feeding. The Galloway is an ancient breed, and at one time was the only one to be found in south-west Scotland. Unlike many others, they have had little influence from other blood added to improve them. For this reason they could probably be seen as Scotland's purest breed of cattle. Originally they were registered in the Black Poll Register together with the Aberdeen-Angus, but had their own separate breed society formed for the management of a Herd Book in 1877. Though early Aberdeen-Angus and Galloway cattle had some similar characteristics, the two breeds have followed considerably different paths. The Aberdeen-Angus is recognised as a swiftly maturing beast that will thrive under intensive conditions, while the Galloway is a perfect hill cow that can exist on the poorest of ground and produce a calf on a low maintenance ration. The Galloway has an outstanding coat of wavy hair that has a soft lining underneath, and due to this can withstand the Scottish climate at its very worst.

John adored the family farm, but in 1963 his family gave it all up and everything was dispersed. He was so upset that he said he had to run away for a while in order to calm down. When he and Maggie eventually moved on, they were forced to sell the dun Galloway heifer and her calf. 'We just had £100, two babies, so I simply coud'nae work, two sheep, three dogs and a wee Singer Gazelle van that I felt was really above oor' station,' Maggie explains. They moved to a farm in the Borders at Greenlaw in Berwickshire, where John was a shepherd, with a hand written contract that reads as follows:

Halliburton Herding

Agreement

£13 per week – cash
Two tons of coal
Twenty stone of oatmeal
Ten hundred weights (cwts) potatoes
Fat ewe in November
2% on nett sale of tup lambs
9/6d per week – two dogs
House and garden

Sadly, they both hated the job and only stayed for six months. The night they left, John wanted to say goodbye to his employer, Mr Elliot. When he went to collect his P45, Mr Elliot's housekeeper told him, 'He doesn't like to say cherio, as he feels you're just like another beast going down the road.' However, John insisted on this farewell. 'After all I had nothing to be ashamed of,' he says.

In 1968 he went to work on Ardnamurchan Estate as farm manager for my stepfather, Michael Thomson. The contract of employment there read as follows:

Agreement

£1400 per year
Car
Coal – 3 tons
Milk – adequate cows will be kept by the estate to provide free milk
Potatoes – grown on the estate, and free

Galloways, Ben Lomond

Keep animals: 12 ewes, 4 ewe hoggs as pack. Must be kept in normal surroundings with other sheep and must get no particular treatment
Holidays
Length of notice: 6 months

'It was really good pay for those days with excellent perks too', John says. Then Ardnamurchan was still largely inaccessible, and most of the animals came in and out by boat from Oban. A hard and exceedingly wet peninsula that lies some 50 miles west of Fort William, it is Britain's most westerly mainland point. The Maxwells adapted well to this complete change of scene, easily fitting into rural West Highland life. 'It was another world altogether where I learnt a great deal about people. At that time there were some 3000 Blackface ewes and 400 native cattle on the estate, including Luings, Highlands, and Galloways. I soon became aware that the Galloway ranged the ground more thoroughly than all the rest, wintered better, and was more prolific. She really is the best converter of rough herbage that I know, and uses the whole hill, even right up to the top. She survives on minimal feeding, and is easily calved. Her calf grows very quickly, and produces beautiful marbled and tasty meat. She may have earned the reputation for being flighty, but frankly, it's like handling a woman. If you talk to her nicely and approach her gently, you will win in the end. The few that do not have a good temperament should be culled out.' He looks at me with that extremely cheeky twinkle and I wonder whether he is referring to women or cows.

'We had a great deal of fun in Ardnamurchan and some of the incidents were hilarious, but I always had to keep on top of the situation. I told the men that I wouldn't tolerate drinking at work, but one of them got very fu' one day, and forgot he had a big piece of new machinery attached to the back of the tractor. He drove straight through the village with it swinging all over the place oblivious to the trail of destruction he left behind him, including the demolition of the petrol pump at the village shop. So he got the push straight away,' explains John who is teetotal.

'During the time we were in Ardnamurchan, streakers were often to be seen on the television during matches and at other sporting events. One of the old crofters, a native Gaelic speaker who spoke only the most basic English, was watching us while we were putting out a Whitebred Shorthorn bull after the winter. Well, the beast had rubbed off lots of his hair while inside and was all pink and fleshy-looking. The old boy saw him and struggled for a wee minute. Suddenly with a huge effort he stuttered, "um, um, um – *by God*, it's a streaker." We all had a damn good laugh and that bull was known as The Streaker ever after.' The Streaker was used to cross with the Galloways to produce the Blue-Grey, a hardy, long-lived hill cow that suited the wet climate of Ardnamurchan very well. The Blue-Grey had been widely produced during the mid-nineteenth century. Its popularity without doubt added to the demand for Galloway cattle, as the female progeny from this cross are prolific milkers, can out-winter in most conditions, and then may be crossed with continental sires to produce an excellent commercial calf. Even today with so many continental beef breeds to the fore, the blue-grey still holds its own and remains a popular cow.

'Another bull was being moved one day and suddenly decided to go into the sea for a paddle. I remember driving past and seeing the stockman sitting on a seaweed-covered rock smoking a fag, waiting for it to come out. By this time you could only see the bull's neck poking out of the water. He stayed in the water for most of the day too. On another occasion a bull was badly sinking in a deep bog – mind you it wasn't a Galloway, they are far too clever for that – so I just rushed up to the pub and asked the 16 folk enjoying themselves to come and help. Some of them were holiday-makers, but they all came willingly, and we managed to pull the poor beast out manually. Everyone had to get on with it out there and muck in. I once asked the District nurse to stitch up my lamb's ear, and she did.'

The Galloways and Highlands in Ardnamurchan were part of the highly successful Mingary-Sunart

White Galloway

Herds when the estate was owned by Boots the Chemist prior to the end of the 1960s. As a child I loved rooting in the sheds and looking through all the old rosettes and prize tickets clinging to the cobwebbed walls, dog-eared and faded, but still legible. There were wooden kists too, full of bristly grooming kits and exciting potions to make hooves and hair gleam, and cover up a multitude of sins. Old sun-curled sepia photographs showed champions proud and square, many lovely Galloways among them.

For John, the Galloway had a large attraction and drew him like a magnet, but Blackface sheep have always been part of his life too. I can still picture him striding across the hill gathering them in ready for clipping. He cut quite a dash, dressed in khaki shorts, the trademark big boots with turned up toes, and a huge hat with a large brim. His eagle eye never missed a trick, and his constant running commentary amused all in his company.

The atmosphere at clipping time was wonderful. The air was thick with the noise of bleating ewes separated temporarily from their lambs. Jostling and pushing, leaping and jumping, each was dragged unwillingly to the clipper. All the sheep were hand clipped in those days. Each clipper had his place, a sharpening stone by his side. Sweat poured from his brow, while back bent double over the struggling beast, he worked on tirelessly. Tar was hastily daubed on to nicks to protect them from the horrors of the flies. Collies snapped at sheep's heels, or lay dozing in the sun-speckled shadows. The noise of shears chinked on rhythmically, followed by the comforting sounds of old ewes reunited with their offspring, whickering softly as they found them in the tightly packed fank.

Then there were the tea stops, when cooled kegs of beer were fetched from our hotel in Kilchoan, and great mountains of sandwiches appeared on trays. Everyone helped with each other's clipping which sped up the process, and added to the fun of the occasion. Nowadays contractors come from as far away as New Zealand and Australia and clip each ewe in seconds. Though this is highly efficient, it provides little 'crack', and it was the crack that was important then. Tales were told, friendships re-established, and new words learnt as yet another ewe sent someone flying into orbit across the shed floor. Kneeling on the ground to roll fleeces, and then leaning into the huge wool sack that was suspended from the great beams of the shed, I became a skilled packer and roller even if I didn't know my seven times table. I returned home stinking of sheep, hands full of gorse prickles, with a huge appetite and a few coppers in my pocket.

When the sack was full, someone small was chosen to climb into it to ensure that the fleeces were tightly packed. I loved jumping about in this huge ovine bouncy castle. After this the bags were stitched up and dragged to a far corner. At the end of the week a mountain of woolsacks lay awaiting collection. Often rain stopped play. The sheep were turned out into an in-by field to await the next dry spell before they could be relieved of their irksome wool.

Once the clipping was accomplished the shed was turned into a dance floor, and young and old joined in for an impromptu ceilidh. Here John excelled himself again, for he is a wonderful fiddler and played long into the night, while everyone jigged unsteadily round the steading, filled with boozy cheer.

John Maxwell is a highly ambitious man, and in 1976 he and Maggie decided to move on from Ardnamurchan. 'It's an awful bonny place, and they're very friendly people, and I wouldn't have changed it to take any other job, but Maggie and I had this life-long ambition to farm on our own,' John explains. 'We moved on to Cashel Farm at Balmaha, on the edge of Loch Lomond, where we were tenants of the National Trust for Scotland. We concentrated on Blackies and Galloways. We had to work with native breeds, and needed to breed all our own replacements as the farm had a bad tick problem, and homebred animals are much more resilient,' John explains. Sheep and deer ticks can cause numerous infections in livestock, and are a constant hazard for hill farmers in many parts of Britain.

'When I left Ardnamurchan your stepfather was extremely good to me, which is something I have never forgotten. He let me buy 10 pure-bred Galloway heifers for the sum of £100 each, and they were worth far more than double that at the time. We did a special deal, and those animals formed the basis for our herd at Cashel. In the early days we enjoyed a great export trade with Germany, but all that stopped in 1990 following the BSE export ban. For a time the high export prices affected the home market as many farmers were unable to compete with the foreign buyers. The Germans treat the Galloway as gourmet food. Why on earth don't we Scots insist on the same thing here? It's all about the *tastability* of beef.' John is adamant that the Galloway produces the best meat of all.

To begin with, the herd at Cashel consisted largely of black Galloways. John, however, had not forgotten his first dun heifer and soon began buying duns again. 'The dun gene pool is very low. I really wanted to specialise in them and try to build up the numbers, but it has been hard to find new blood, so we have had to look to Canada where many good animals have gone in the past. When we eventually moved on from Cashel and bought the Jaw Farm, my son Duncan and I decided that he would keep the black Galloways on his farm, Blairvockie, on the side of Ben Lomond, while I would take all the duns, and concentrate purely on them. It has worked very well. Both my sons Duncan and John are good farmers, and my daughter Joy knows her stuff too. We all help one another,' John says.

'At Blairvockie all the animals are wintered outside, but here at the Jaw, we do bring some of the cattle in, especially as our winters seem to be getting wetter and wetter. It is only because they poach up the ground so badly if we don't. When they come in they have to be clipped, otherwise with so

Peacock butterfly

Small tortoiseshell butterfly

much hair they would sweat up too much, and probably get pneumonia. Hair is of course a breed characteristic. They are such attractive cattle don't you think? I love judging them at shows and I really don't heed whose watching, or whose holding them. Unlike some judges, I can then justify my decision correctly.'

John is particularly pleased with a young dun bull he has recently sold that was still waiting to leave the farm when we visited. 'Jaw Hiawatha is a good example of the breed, and has a great nature too. I like the cows to be big enough to milk well, and large enough to produce a cross-bred calf too. Duncan and I both finish our own steers off from grass and they are usually sent to the abattoir at between 26–28 months of age. This has to be the main commercial venture, but I have had to do a variety of additional things to keep the farm going. For many years I was an agent for Poldenvale, a company that

Kingfisher, Drumwalt Farm, Old Place of Mochrum

manufacture gates and equipment for livestock-handling systems. I was also an agent for a fertiliser company. However, without question the farm has always come first, the other things just help to keep us.' There is that typical grin again from under his broad brimmed hat, and a few parting, teasing comments. An adaptable and intelligent man, John does not need to tell me that he has never been one to rest on his laurels.

The Galloway is a stunning breed. I ask John about the wonderful belted varieties. 'That's something I know nothing about,' he says 'but the person you should talk to is Miss Flora Stuart from Port William. She's a delightful wee body, and will know as much about them as anyone.'

The road to The Old Place of Mochrum in deepest Galloway is long. From Scotland's Booktown, Wigtown, narrow roads run down the Machars, which is the name given to this far-flung corner of Wigtownshire. In spring, summer, and early autumn the hedgerows are filled with berries and wild-flowers, and as few agricultural pesticides are used, butterflies are still a common sight. I remember cycling with my son round the area in early September, and seeing peacock butterflies lazily flitting along the verges, settling on bramble leaves that were just beginning to turn in colour, and small tortoiseshells on purple spear thistle, while thistledown and rose-bay willow-herb fluff floated along the ditches, back lit by Indian summer light. The Galloway landscape is patterned with dry stone dykes, and as you travel nearer to the sea the trees become stunted and windblown, sculpted by the salty winds that drive in off the western seaboard. Knobbly hawthorns are decorated with soft green lichens, and ivy grows in abundance, often forming a thick curled stem almost as dense as a small tree trunk. This part of Galloway has a timelessness that seems to make the pace of life so much slower. Though the fields are mostly small in size, the moorland stretches for miles, edged with scrub wood-land, and fringed with bog cotton and flag iris. Kestrels hover by the grey dykes, and hen harriers are frequently seen quartering the heather-clad ground. A sparrowhawk may dash low over the road and into the woods, while otters frequent the abundant water-courses. This is the childhood haunt of the well-known author Gavin Maxwell, and Elrig, his old family home close to the sea, is surrounded by wild landscapes that clearly inspired his beautiful writing.

Many parts of Galloway are densely forested. In recent years barn owl numbers have risen in the area, and on the forest edges these ghostly spectres emerge at dawn and dusk to hunt for voles in the misted shadows. But perhaps the most dramatic sight of all is the stunning herds of Belted Galloway cattle, animals that blend perfectly with this largely unspoilt corner of south-west Scotland.

They paint an eye-catching picture, and for this reason must surely be one of the most beautiful breeds of cattle. The Belted Galloway is now formally recognised as a sub-species of the Galloway. It has its own separate breed society, and a Herd Book that began registering animals in 1922. Though the Beltie has essentially the same origins as the Galloway, the white marking is thought to have been brought in by an infusion of Dutch blood, from sheeted cattle in Holland sometime during the seventeenth century. The black Beltie is the most common. At the Old Place of Mochrum however, there are dun and red Belties too, as well as pure white Galloways that are attractively marked with clearly defined black etched round their eyes, black hairy ears, and a black nose. The riggit with a dark or red ground colour, and a white line and greyish markings down its back, was thought to be extinct. They had indeed almost died out, when a few years ago a white cow and bull produced a beautifully marked riggit bull. Soon after this Flora Stuart discovered that several other cows had produced riggits too. There are a few riggits in Germany, but this variety is unusual and seldom seen.

The Stuart family have been famed for their Belties for generations, and their herd is one of the few left in the country that formed the foundation of the breed for the Belted Galloway Herd Book. Flora Stuart now runs the estate. Her father, Lord David Stuart, a son of the Fourth Marquis of Bute, spent 15 years researching his book, *An Illustrated History of the Belted Galloway*. His whole life was devoted largely to Belted Galloways, and like her father and grandfather before her, Flora has been passionate about them all her life. She is President of the Belted Galloway Cattle Society, and has done a great deal to promote the breed both at home and abroad.

There are parts of the Old Place of Mochrum that date back to 1400. Drinking water is still drawn from a pump in the cobbled courtyard, being of a much purer quality, and milk for the house is provided by Beauty, an Ayrshire cow. Flora claims that both the water and the milk are unrivalled in quality being free from additives and non-pasteurised. The beautiful ancient castle is surrounded by lichen-covered trees, and in the orchard gnarled and laden crab apple trees provide shelter for a small flock of coloured Shetland sheep, and a feast for the birds in winter. Flora uses the Shetland's soft wool for spinning. Black Rock hens rush up to the car as you arrive, clucking soothingly in the hope of a morsel of food. Close to the farm the Mochrum Lochs provide an ideal habitat for goosanders, mergansers, and golden-eye. Large pike are found in the lochs, but voraciously devour most other fish. Greylag geese, snipe and curlews nest on the moorland's raised blanket bogs, and this is one of the few inland-nesting sites for cormorants.

Behind every traditional Galloway dyke, another group of Belted Galloways is revealed. The dun and red Belties are equally as eye-catching as the more common black, although the latter still remains the most popular. Many of the animals are selected for showing during the autumn of the previous year. Thoroughly prepared for the show ring throughout the winter, they are halter-trained and given leading practice. They are shown in leather halters. The animal's first outings will be to small local shows before going on to bigger venues such as the Royal Highland Show at Ingliston, the Royal Show at Stoneleigh, and the Great Yorkshire Show. One black Beltie, Mochrum Nina, succeeded in winning Junior Championship at all three shows in 2000. The bulls have rings fitted in their noses when they are just over a year old for ease of handling. Flora loves the showing side of farming and does much of this herself, together with the stockman, Tommy Frame, whose father worked on the estate before him for over 40 years. Though their achievements and championships are many, and one of their cows, Mochrum Kestrel, was the first Belted Galloway to win a Champion of Champions award, shy modesty is very much a part of Flora Stuart. While her dedication and total devotion to the Belted Galloway has clearly been the most important aspect of her life, you will not hear her speak of her immense achievements within the breed.

The cows calve in the autumn in order to save any conflict with lambing, for as well as the small flock of Shetland sheep, the estate has a substantial flock of Blackface ewes. The bull calves are castrated in

(opposite) Belties, Fell Loch, Mochrum

the spring, and are usually sold at just under a year old. The heifers that are not being retained for breeding are sold when they are about two years old. Flora knows every animal, its calving record, bloodlines and history, and explains their background to me as she takes me round. It is early November and the bracken has turned a rich bronze. Many of the cows have just calved. The youngsters have formed a large creche and skip about on the moor, with the Galloway hills blue in the distance behind them. In places the mud is up to the top of our boots, as it has been an immensely wet autumn. The Galloway has earned a reputation for a fiery temperament, and there is no doubt that when a heifer is newly calved, care and respect must be shown. However, Flora knows her animals and is content to wander quietly among them even with their new calves, although she does carry a stick as a sensible precaution. It is obvious that she has a great rapport with her beasts.

Mochrum Kingfisher, a black belted bull stands and watches us from a knoll in the next door field. He is a magnificent animal with a thick curled mossy coat that keeps out the Galloway dampness. Like most bulls, the Belties occasionally break out. With other breeds, resulting offspring can be of a nondescript colour. However, there is one drawback with the Beltie, for they nearly always leave belted progeny behind them. It must be very hard to deny that it was your bull that intruded on the neighbour's heifers, when the resulting offspring have that unmistakable tell-tale white middle.

In the final paragraph of his book, *An Illustrated History of Belted Cattle*, Lord David Stuart closes by saying:

> As I draw nearer to the end of the pilgrimage of life, as the shadows grow longer and this world creeps on its petty pace, I am more able to ponder how blest I have been in my time. Not least among those blessings I count the pleasure I derive from the great love my daughter has for her inheritance and the knowledge I have that she will hold on to her Belties as long as she is able.

Having spent precious time in her company, and shared briefly a part of the Old Place of Mochrum, I realise that the Belted Galloway owes the Stuarts a very great deal, for no family could have been more truly dedicated to their survival.

Kingfisher

CHAPTER THREE

Three-In-One – The Infamous Blackie

MANY areas of upland Britain can be bleak and inhospitable, growing little of nutritious value for animals. However, this is the haunt of our most numerous breed of sheep, the Blackface. Its origins have drifted into the annals of time and are impossible to trace. Adept at withstanding the harshest of weather, the Blackie can scramble goat-like on to inaccessible ledges in order to obtain a succulent blade of grass, and is perfectly suited to the rigours of mountain existence.

The life of sheep on such hard ground is short as their teeth wear quickly on heather and tough plants and grasses. After they have bred usually four crops of lambs, they will be sold in the market as 'cast' ewes and bought by lowland farmers to cross with other breeds of sheep. The Blackface has been widely crossed with the Blue-faced Leicester and the Border Leicester to produce popular and prolific breeding sheep, the 'Mule' and the 'Grey-face'.

It was 'cast' Blackface ewes that taught me a great deal about ovine behaviour, and quickly dispelled the myth that sheep are stupid. The competitors on the awe-inspiring programme, 'One Man and his Dog', have clearly never been given the belligerent devils that we brought home from market. Dexterity, calmness and rapport between man and beast was always very apparent on the television programme. But our Blackies were frequently more stubborn than a mule and had a total disregard for the collies. Handling them sometimes left us anything but calm. One hardy, wily old ewe became skilled at backing the collie up against the electric fence, treating her to manic head-butts at the front end while the poor dog was receiving high voltage shocks to her rear end. Yelping in terror, the dog would flee to the sanctity of her kennel to cower, and it mattered little which tune we whistled, for, after such humiliating treatment she would not emerge for the rest of the day.

The Blackie is a skilled escapologist, a veritable 'Horned Houdini'. Some can leap as high as a show

Feeding ewes, Doldy Farm

jumper while others merely hurl themselves straight through fences and gates. No one likes receiving an injection, and sheep are no exception. Struggling and furious, they frequently send the syringe into the wrong destination as they battle to evade it, and it wings its way torpedo-like into the thigh of the shepherd. While the horns of the Blackie can act as useful handle bars as they are manoeuvred round the fank, they also catch in the pockets of clothing, either leaving you without somewhere to put things, or, worse still, pulling you into the sheep dipper with them. Sheep can bring out the worst in you. The Blackie can be as wild as a hawk. But it is these very traits which make her so unique.

Over the years three distinctive types of Blackface have evolved to suit specific areas. These types have been named after the market towns where they were sold and are traditionally the Perth, Lanark, and Newton Stewart. All three variations, and several other localised types that are perhaps harder to classify, are horned and have black, or black and white, faces and legs. Fashion has without doubt played a significant part in the evolution of the breed and will always be subject to much conjecture and controversy.

As Dave Nicol from Doldy Farm in Glen Isla, a well-known breeder of the Perth Blackie, astutely said: 'It's very hard to keep abreast of the situation when they keep on moving the goal posts.' The Perth type is a large sheep with quite long legs, plenty of length to its body and beautiful, heavy, long wool. With their large angular framed carcass, Perth Blackie lambs tend to be slightly slower to fatten and have to be kept for a longer period before they are sold. The horns of the Perth Blackie tend to come out from its head at a distinctive angle, falling back away from the face.

Dave has farmed at Doldy for 30 years, and has lived in the beautiful Angus Glens all his life. He previously shepherded at Glen Uig, but started at Glen Moy, where his father, who taught him much of what he knows about sheep, was head shepherd.

Crisply sparkling snow was covering the ground on our first visit to Doldy. A hay hake surrounded by ewes with bright coloured paint on the back of their necks, corresponding to the raddles of the tups that they had been with, created a peaceful winter scene. Next day the picture had changed totally as torrential rain fell making the yard at Doldy an icy skating rink topped with a layer of water.

Many of the tups were housed for the winter in the old-fashioned steading. The buildings were airy despite the low cloud that hung like a melancholy veil over the whole area, and some of Dave's show hopefuls were peacefully cudding in their pens. Dave is very keen on showing and has a huge selection of wonderful photographs of champions dating back many years. 'When they are born all lambs are like children, perfect little angels.' Then Jan, Dave's wife, quickly adds, 'as soon as they are born Dave is convinced another champion has hit the ground and he's sure it's a real winner. It's not till later on that all the faults develop.' Doldy Blackfaces have won at the Highland Show many times, and Dave has also judged Scotland's most prestigious show himself. 'It is a job I look forward to, but I know how hard it is to keep everyone happy.'

Doldy carries 650 ewes, and one third of these are now the Lanark type. 'The Irish used to be very keen on the Perth type, but the Lanark sheep are gradually taking over, and with no money in wool, the longer fleece produced by the Perth sheep is not so important. The fleeces used to be sold for the Italian mattress trade, but not any more. Its very hard to keep up with the constantly changing market.' Despite this, Doldy's main flock still remains the more traditional Perth type. Originally the curvaceous horns of the tups were filed, shaped, and polished. Now there is little interference, although many of them will have a small hole drilled in their points so that a sprung wire can hold them in place to stop them from growing too close to the face and causing problems.

The Lanark Blackface has a smaller body and wool of medium length. It is more easily finished, and therefore does not take so long to fatten. Currently, market trends have swung towards the production of a lighter carcass for which the Lanark type is ideal. The horns of the Lanark tend to rise up and out of the head. Even to the uninitiated, when the two types are side by side it is easy to see

Doldy tup

their differences, particularly in their horns and differing fleeces. Lanark sheep have had some Newton Stewart blood introduced into them to give them better milking ability and hardiness.

Farming at Connachan, 1000 feet above sea level in the Sma' Glen, near Crieff, Neil McCall Smith and his daughter Mary are well-known figures in Blackface sheep circles.

Neil admits, 'my father didn't know much about sheep when he first came to Connachan Farm just as a temporary measure, but he was a very good manager. There have been McCall Smiths at Connachan ever since. My middle name, Logan, was the surname of father's bank manager. I hope it helped. My father bought our original stock from Abercairney Estate, and they had taken them over from the previous tenants. We paid £1900 and I still have the receipt. £70 of that was deferred as we were on a very tight budget.

'In father's day the sheep were looked after to keep them alive, and were never fed in the winter, which I think is quite wrong. Now we feed them for production, and we also give them lots of minerals. Lambs obviously bring in money and we try to sell our sheep on their hardiness. The land dictates the type of sheep it can carry. Much of the land in Scotland will carry a big ewe, but further west smaller sheep are more suitable. Areas like the Lammermuirs have the best land because it's just soil there. The Blackface ewe must be able to not only exist, but also thrive on the ground, and produce an average of one and a half lambs per annum. At the end of four crops she should ideally be sold on to softer ground, with a full mouth, and her milking machinery in order. You must always pick the right sheep for your type of ground, and our farm suits the Lanark Blackie well. You'll often hear shepherds in the market saying, "that farm carries a guid big yow," or perhaps, "ah well, oor ground'll no carry an awfy big yow." It really does depend on where you farm. However, the Blackie has the ability to put condition on its back on an indigenous pasture. With the farm being quite high we frankly don't want too many lambs. One year we had 500 pairs of twins; well, it was too many for 2,000 acres.' Neil laughs.

About 50 years ago the McCall Smiths changed from the Perth Blackface to the Lanark. Their sheep were demanding too much attention. In bad winters they had to have their wool trimmed to avoid the snow balling up on the fleece. 'Sometimes in spring the lambs found it hard to suckle because the wool had become tangled up round the ewe's udders, so we realised that our hard ground was more suited to a smaller sheep with shorter stapled wool (staples refer to the length and type of the wool). Perth sheep were good from a financial point of view when there was money to be made in wool, because of their big heavy fleeces, but not now. These days, it actually costs us to clip our sheep, and we can hardly give the fleeces away. It's tragic.

'Forty years ago we used to put an eggcupful of whale oil into the sheep dip to help the fleeces from becoming matted, but my wife made me stop that because of the threat it posed to the poor whales. Now you have to have a certificate to be allowed to dip sheep because of all the chemicals present in the dips. Everything has changed.' Fields of brilliant orange and brown tups are not an unusual sight. Some farmers have even used florescent pink to 'bloom dip' their animals. This brings new meaning to seeing pink elephants. These dips, so out of keeping with Scotland's dramatic scenery, are purely for appearance and are used to make the tups stand out. 'It's purely cosmetic; the Lanark boys like to colour everything. Years ago, we used to take some clay from the back of the farm and colour our tups with that, but it was nasty stuff, the colour of digestive biscuit, and washed out in the first rain storm.

'Blackface sheep breeders are old-fashioned. Things are ingrained into the people as much as they are into the sheep. We do not have a Herd Book, but if you're a sheep person you can tell if the lambs are from a particular tup. The sheep have to sell themselves – we're not like Hoover salesmen who are so pushy. One year we bought a new tup but we didn't like the lambs he left. He cost us £8000, but despite the money we sent him to the slaughterhouse. We make mistakes, it would be so easy if we didn't, and we'd be millionaires.'

1997 was an outstanding year for Connachan. A tup lamb sold for an astronomical price. 'People

Ewe and lambs with blackcock

Connachan tups

thought it was a wangle, and poor Mary was very embarrassed. She couldn't believe it. In our estimation, I think we have a better one this year. Sadly there is simply no incentive now to invest in sheep, hill farming is in an awful state.' Prices of Blackface tups are steeped in mystery and speculation, and the Connachan sheep always seem to make a steady trade.

'Our ewes go to the tup in November after they have been flushed on fresh grass. This ensures that they are in good condition and will be most likely to conceive. Clean grass is very important to sheep. We have 35 suckler cows to keep our pastures in order. They have their calves when the ewes are lambing, in April. We try to stagger our lambing a little to make life easier. In father's day they had four men to do the work that we do now. Then the ewes were in much poorer condition and sometimes there were heavy losses. Some wet, sleety days the consequences were dreadful and I remember seeing barrow loads of dead lambs. If you are assisting sheep to lamb, it is helpful to have nice small hands, not great muckle peasant's hands like mine. We have been lucky and have never really lost many lambs to foxes. But you mustn't leave anything dead lying about for a fox to acquire a taste for it. I'd much rather they had a taste for grouse than for lamb, but of course my landlord wouldn't agree.

'I used to help a great deal on the farm when I was a boy. I was twice as efficient and co-operative when it was time to go back to school, because I hated it, and sometimes I might be kept off to work on the farm. As well as the shepherds, we had an old boy who dug the garden and checked the drains for my mother. Mother was a city girl. She used to take the mare and the spring cart up to Crieff and hand the horse over to the policeman whilst she did her shopping.

Neil McCall Smith

'I am reputed to be very mischievous and I think it's a fair judgement. Our local schoolmaster was always drunk, so that was hopeless, and my parents hired a governess. One day when she was teaching me the piano, I got fed up with all the "do, re, mi's", and slammed the lid on her fingers. There is simply no music in me. I'm what father called "timmer-tuned." After that misdemeanour, I was sent away to boarding school. I used to put oil into the inkwells. "Was that you, McCall Smith?" the master would ask, and I would reply, "you can't prove it," and of course they never did. I worshipped some of the masters, but there were others that I despised. Once, when I got a good caning, I said, "thank you very much, Sir, good morning." "Indolent boy," the furious master mumbled. College was totally different, as it involved interesting things, like how to grow grass, and cattle husbandry.

'When I was ten years old, we went to a fete at Abercairney. I was immaculately dressed up in a complete sailor suit. I really thought I was the "Cat's Whiskers." The Laird was held in great esteem and Father said, "Now, Neil, if the Laird's wife's Pekinese lifts its leg on your trousers, you must say 'good little doggy' and smile politely." If you were on the wrong side of the Laird or the keeper you were in real trouble.'

When Neil returned home from College in the 1930s, his father asked him what he was going to do. One of their shepherds was leaving. 'You can have his job, but I can't afford to pay you,' his father offered. As well as working hard on the farm, Neil kept 300 White Leghorn hens. 'They were kept for the hen trade, for eggs, and not for meat. The hen house was lit with a nice soft light to keep them laying, and a Japanese person came to sex them.'

Neil McCall Smith is renowned for his stories, and time flies as we sit in the farm kitchen. A tabby cat appears at the window carrying a young rabbit, the kettle simmers on the Aga, and an ancient collie snores in its basket under the table.'So much has changed for the hill farmer, we are a dying breed. I was conceived, born, and still sleep in the same bed. God and the landlord willing, I'll die there too.'

At Cuil Farm, Palnure, Newton Stewart, Graham McClymont and his family are a force to be reckoned with as far as the Newton Stewart Blackface is concerned. This third type of Blackie is basically the same as the Lanark, although it has a few conformation differences and a slightly different character. It produces a great deal of milk. Its body is closer coupled with a shorter fleece and is ideally suited to the high rainfall of the south-west of Scotland.

Cuil has a fine hilltop position overlooking the snaking Cree Estuary where the mud flats change the landscape and the reed beds provide a haven for wildfowl. We received a great welcome at the farm by Graham and his wife Christine. Graham's father came to Cuil in 1938, and the family has since built up a considerable reputation with their flock of 1000 Blackies.

The walls of the sitting room are closely packed with framed black and white portraits of their sheep. From these it is evident that the look of the animals is constantly changing. 'It's all fashion,' laughs Christine with a twinkle, 'Half the time I don't think Blackface breeders know what they are wanting themselves.' Of course the men in the family would not agree to this comment, but are clearly used to her sense of humour.

I ask Graham and his son Colin to point out the best sheep on the wall, and they point to a tup with a greying face, short, tight wool, and horns not dissimilar to a mouflon. This is clearly the Blackie of the future and bears little resemblance to the long-woolled beast of years ago, with its lengthy body and swept back horns. 'When I started farming I soon realised that it's no good having the wrong sheep for the ground. The sheep must thrive on your ground, and ours are thrifty and have barer skins than the other types, and they are much better milkers. The Newton Stewart Blackie is the hardy one.' Graham is adamant about this.

'Years ago we were heavily penalised if our sheep had any black in their wool, but now it does not matter much because the wool is so valueless.' From the fine gallery on the walls, tups with 'bell-brows' – a white mark on the forehead and badger-like 'brockie' markings stare down, reminding us of the changes. But the modern Newton Stewart Blackie's face often appears grizzled and grey about the muzzle, and even the stance seems to have altered.

The traditional Galloway dykes have greatly shaped this beautiful landscape, and Cuil has many fine examples that provide wonderful shelter in every windswept field, and lend timeless character to the farm. Many of them are ancient but they are all beautifully maintained. At the sheep'buchts' high above the farm an icy blast blows off the Lamachan Hills. The clarity of light in Scotland can be quite unique. February's sharp luminosity paints the stretching panorama far below us as we lean on the huge lichen covered dykes and listen to the haunting calls of the curlews. A buzzard cruises the thermals, and a gnarled hawthorn, bent almost double, has clung to the dyke-side for solace from the wind, to

Cuil, Newton Stewart, 9th Feb 99

little avail. Graham and Colin show us some ewes and gimmers (two-year old ewes) which they consider are typical examples of the Newton Stewart Blackie. Their wool blows in the sharp wind, and they stand by the dyke stamping their feet at the slinking collie, showing the forceful character so typical of the Blackie. As we blow on our hands and pull our coat collars closer, the clear day carries snow on the wind. However, Christine's hearty soup soon warms us as we return to the kitchen to talk sheep again.

Many of the McClymonts' tups once travelled by train to Dalmally, near Oban, to be sold in the market there. 'In 1953 I remember there were 200 tups on the train. There's a lot of luck in this tup business and when you are in the market you can see in a flash what you want and, I suppose, often you just pay. But tups can die too and you might end up losing a great deal,' Graham explains.

Perfectly at home in the wildest of Scottish landscapes, thriving in the worst of our erratic climate, hardiness is the outstanding characteristic of this beautiful sheep. Hardiness is also clearly evident in those who make their living with the infamous Blackie. As Robbie Harrison, a shepherd from Glenartney, Perthshire, who knows and understands the Blackie well, said: 'Blackies may vary greatly in size, colour, and skin type, but they are all bloody awkward. Do they actually discuss at night which of them is going to be dead by morning for no apparent reason? And why do they think of a five-foot high dyke with three wires on the top as a challenge? This is because they are truly great sheep: wild, thrawn, tough, canny, hardy, and sadly, greatly undervalued.'

(opposite) Newton Stewart ewes, Cuil Farm

CHAPTER FOUR

Dod Orcheston – The Stickmaker

DOD ORCHESTON of Kirriemuir, Angus, has been making sticks using Blackface tup horns since he was a youth living in a tiny bothy in Glen Prosen. The long dark nights of a Highland winter provided him with the perfect opportunity to learn the craft from old shepherd Fred Mitchell when they were together at Pearsie. 'That was when I was 18 years old and I'm 78 now and have been whittling awa' at horns ever since. I had nae idea that it would snowball into what it is now. There are sticks o' mine all over the world. At the beginning it was just a way of passing the time after work and keeping mysel' oot o'mischief!' Dod Orcheston has a magnetic twinkle and wonderful sense of humour that often veers on the wicked side. Stepping inside his bothy workshop is like turning back the clock. Once the swollen, warped door creaks shut behind us, time seems to stand still.

The embers of a fire glow in the black basket fireplace sending patches of dancing light on to the ancient flaking stone work. A spider's paradise, dusty cobwebs hang in festoons in every cranny like wispy mist. A bookcase houses a selection of yellowing volumes and faded, dog-eared machinery manuals. Gnarled and weather-beaten horns of all shapes and sizes hang between horse harness, tools, and pots and jars of potions vital to Dod's craft. Above the fireplace there are cruisie lamps and candles, tins of polish and beeswax. Hardened amber-coloured beads of varnish have oozed stickily from a battered tin, like sap from a pine tree. A faint aroma of singed horn reminiscent of childhood visits to the village smithy adds to the atmosphere. There is no intrusion of a telephone and nothing to indicate our modern era.

'I like to mak' sticks and sit and look at them. I never sell them to shops, the commission bugs me, they are ay' needin' mair for handing them over the bloody counter. I like to sell them frae hame. Recently I was sitting in the pub haein' a pint when some folks frae Los Angeles walked in. They were

(above) A future Crook?

looking for me, but I did'na let on. Once they knew it was me they asked if I had ony sticks to sell. I micht, I said, but you'll hae to wait till I've finished my pint for I'm no leaving it. They bought a crosier for their son who is a minister, and several others. Sticks are important, they can be as guid as a third leg. When I had my hip done, I was issued wi' sticks by the hospital, but I wouldn'a even hae thrown them at the dug. Those weren't sticks.'

It takes approximately 50 hours for Dod to complete one of his beautiful sticks. The more ornate ones will take a few hours longer. The traditional 'Cromack', which has a carved thistle, is his favourite. They are unpainted, demonstrating the exquisite tones within each individual horn. Every delicately carved thistle varies slightly. A Cromack is unlike a walking stick and consists of a straight hazel stick of more than 4 foot 6 inches in length, a tup's horn and a ferrule. The ferrule is the piece that holds the horn on the stick and is cut from a cow's horn. Usually Dod will cut enough material for two ferrules from each horn. The remainder is never wasted and is used to make spoons, napkin rings, and shoe-horns. 'I always use a shoe horn. When I was wee, it was a shoe-horn or a bang on the lug.' He also makes brooches of Scottie dogs, horse-shoes and thistles. 'I never throw onything awa', I'm a noted hoarder, jist a richt magpie. This place is jist a bloody glory hole, I'd hate to be my daughter after I'm awa', she's going to have to clear all this up.'

The sticks themselves are mostly made of hazel, but holly and blackthorn can also be used.'The time to cut hazel is when you see it. If you don't some other body will. I used to get sticks oot o' the Den O' Airlie, and then one day I met three men staggering along weighed doon wi' aboot 300 hazels. I said, d'ye ken fits wrang with you – bloody greed. They ruined the den for a'body else. Normally I try to cut hazel sticks between November and March when the sap is oot.' Once cut, the sticks will have to be left to dry for at least a year. All Dod's sticks are surprisingly straight. When I ask how he achieves this, he replies, 'Ony stick can be straightened with a little patience, perseverance, and know how, but how I do it is my own secret.'

Blackface rams' horns are becoming increasingly hard to come by. 'Guid horns are like hens' teeth. Nowadays many of the hardy hill sheep are fed and pampered, and the horn ends up being far too saft. You win some, you lose some, you tak fit you can, and be very thankful. However, I have never bought a horn in my life. Folk that hae horns are aye needin' a stick. They'll get a guid stick too, so it works both ways. I'm lucky to hae guid friends who supply me.' Many of his horns come from Glenlivet. The tups there produce large, hard horns which do very well when they are put in the fire, which is the first stage and which cleans up the horn, removing any unwanted smell. Dod instinctively knows exactly how long to leave each horn in the fire. Too long can ruin them altogether. Once out of the heat, the horn is put into a vice and shaped or bent, using gentle pressure, before it is left 'to set.' It will then be left to cool.'Once it's set you can then decide fit you are going to do wi' it.' It is at this stage that Dod will see the potential for a certain creature, or perhaps a thistle, and an idea for a design will be born.

Every stick is totally individual and depends on the soft hues that emerge once the horn has been whittled, rasped and polished. Dod does not believe in adding any pieces on to a horn, and likes his sticks to be as natural as possible. Some will be painted. The array of animals, fish and birds that have evolved on the lovely curvy handles of the sticks is awe-inspiring. Brown trout appear as if fresh from the water, their lithe bodies sinuous and gleaming, an owl sits on a book, and there are Clydesdale heads and stealthy working collies lovingly carved from each horn.

'I've had some unusual requests, one tattie Merchant from Wales asked me to mak' a tattie on his stick. To me it jist looked like a lump, but the artist made a braw job o' it and the man was highly delighted. I'm not an artist, so I find other bodies to paint the beasties once I hae carved them. They mostly use enamel paint, but I still prefer those of natural horn colour, the "Cromacks". It's important not to lose that old Scots word.'

Many stick-makers take a pride in competing in shows. 'To me it's jist a cause of nothing but

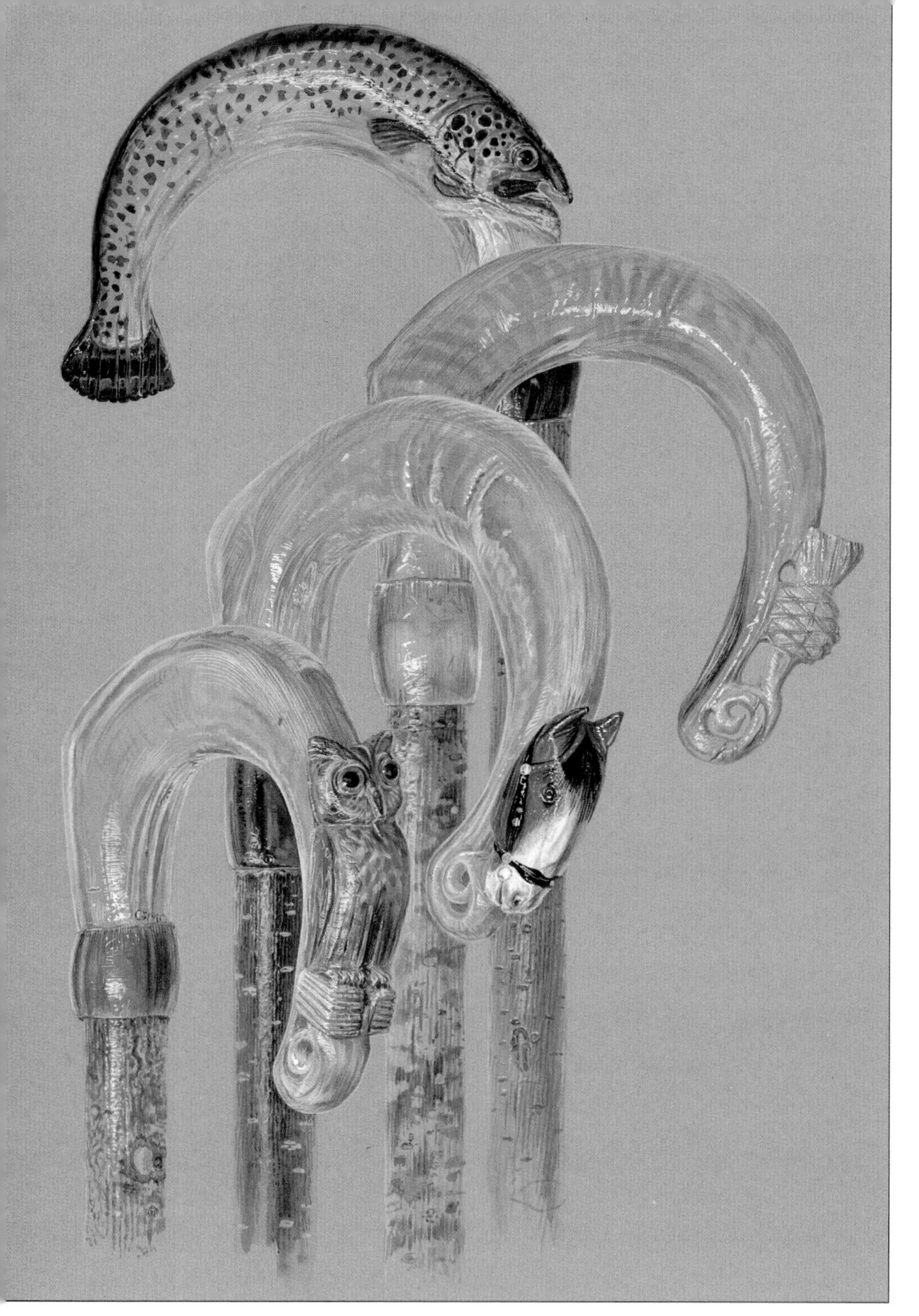

trouble. Folk get so jealous. Sometimes I am asked to judge stick competitions and I can assure you that it's even worse.'

All the finer work is carried out in his workshop in the house, where a vast selection of files and small hand tools play a vital role. After whittling and carving the horn to the required shape, there are endless polishing stages using fine emery paper, wire wool and sometimes Brasso, which is a mild abrasive. Beeswax will add the final touch, and lastly the horn will be coated with clear polyurethene varnish to protect it.

Many, many hours have been passed fleetingly in the workshop, time accompanied by a pipe of 'black bogie roll' and more than the occasional tipple. 'I used to like a smoke but have now given up and just hae my dram. In my youth, if you were a plooman or a fairmer and you had a pipe, a moll and a pair o' horse, you thought you were a man.'

So much has altered since Dod's youth. Sitting before his bothy fire, he remembers everything as if it were yesterday. A great lover of poetry and the works of Robert Burns, his fluent recital of 'Tam O'Shanter', amongst other long poems, is quite remarkable. Particularly when he tells me that he left school at the tender age of 13. He played truant as often as he could and landed himself in many a tight spot. On one memorable occasion he was caught smoking in the school boiler house and brought up before the dominie. '"Yes, boys, you were caught today, so who's got the fags?" I had to admit it was me, and the whole incident cost me a bloody fortune; instead of the belt coming out, the dominie hands me the fags and I had to smoke all eight of them. That was sixpence worth, what a waste. You know, he was a cracker of a dominie.'

Dod's beautifully made sticks show a dedication to a lifetime of craftsmanship. They have brought pleasure to many, and are testimony to his skill as a carver. Importantly, he says, they have brought him many good friends. However, there is now a fearful possibility that bureaucrats sitting unawares in Brussels may pass ridiculous laws that will stop him and many others like him from using horns. Sheep's heads may have to be burnt and it will not be legal to remove the horns from them. We have all become crazily over-careful and health conscious, and far madder than any cow. If we do allow such ridiculous laws to dominate us we will lose many vital facets of our valuable heritage. 'These horns will be around long after I'm in my box, it's just a load of bloody nonsense to stop us from using tup horns. A shepherd without a crook is like bacon without eggs.'

Dod Orcheston, Kirriemuir
15th October 1998

CHAPTER FIVE

Stalwarts of the Beef Shorthorn

THE larks were flying high in the sky on a glorious day in May as I drove up the long farm road to Fingask, Dairsie, Fife. The air was heavy with the sweet coconut scent of gorse as swallows busily patched up their old nests, collecting dry mud from a nearby gateway. Up on the hill beside the terracotta-coloured pantiled roofs of the traditional Fife farm buildings, a group of Shorthorn cattle lay cudding in the sun.

During the 1830s, the Beef Shorthorn was greatly developed, and its qualities as a producer of prime beef were enhanced and improved by Amos Cruickshank from Aberdeenshire whose herd of cattle ran from 1840–1890. At that time it was frequently known as the Scotch Shorthorn, and was a popular and highly sought animal in England as well as north of the border.

Willie McGowan, now well into his eighties, has loyally dedicated much of his life to the Beef Shorthorn. And despite the temptations of bigger, foreign bovines, he has not for one moment thought to change them for any other breed of cattle on his Fife farm. Indeed, the breed owes a great deal to his dedication, and it is his involvement with the Beef Shorthorn that has rightly earned him his much-coveted MBE. Amongst Willie's many achievements, he was also President of the Shorthorn Breed Society.

As we sit drinking coffee in his kitchen surrounded by a lifetime of photographs, cups and rosettes adorning his walls, he wryly laughs as I ask him about his trip to Holyrood Palace to collect his award. 'Well, frankly it was a wee bit disappointing. It was fully a twa' hour job hanging aboot. Surely when you meet the Queen, well they could have at least gien' you a cup o' tea with her couldn't they?' This is typical of Willie whose dry, quiet humour is ever apparent.

It was almost 50 years ago that the first Beef Shorthorns arrived at Fingask. When Willie returned

(above) Swallows

from a horse sale in Lanark, he found six cows standing in the byre, bought by his brother in Aberdeen. Shortly after this they acquired their first bull, Brigadier General, at the Perth Bull Sales, for the bargain price of 340 guineas. However, later on the two brothers discovered that something was amiss when the insurance company informed them that the bull was insured for two thousand guineas. A little cosmetic adjustment at the sale had concealed the fact that Brigadier General had a black nose – an unacceptable defect in the breed at that time. However, the Brigadier left some good calves, although Willie admits one or two of them did inherit their father's nose.

I first met Willie at the famous Perth Bull Sales on a sad day when Shorthorn prices had dropped to an all-time low. He stood in the ring with a gleaming roan-coloured bull that was wearing its traditional leather and brass head-collar, as auctioneer David Leggat tried to drum up enthusiasm. For Willie this was a far cry from the days when there would have been a whole week dedicated to sales of Beef Shorthorns, with perhaps over 300 bulls entered. On this day, he only had three for sale. Competition with foreign imposters has been hard to contend with, as continental beef breeds have not only completely overtaken from Britain's native animals, but have also been highly in vogue. Even the venue for the Bull Sales has changed. Now like so many markets it has been moved to a site on the outskirts of the town, and due to this, it has sadly lost a great deal of its atmosphere. In its heyday, the Mart was situated right in the town centre, and much of the judging took place in the streets. It was not unusual for the odd beast or two to take a wee hurl round the shops when they occasionally broke out. 'In those days we won a great deal, but we didn't get drunk on it,' explains Willie. Others would certainly not be so able to make this statement as the Bull Sales were usually a great place for a drinking spree. The barman of the Station Hotel in Perth used to claim that even more whisky was consumed during Shorthorn week than during the week of the Aberdeen-Angus sales.

Another stalwart of the farming community who has done a vast amount to promote Scotland's native breeds, is Captain Ben Coutts, of Woodburn, Crieff. An author, judge, broadcaster, and farming commentator, to list but a few of his achievements, he is famed throughout Scotland, and a good deal further afield, for his knowledge and staunch support for not only the Aberdeen-Angus and the Highland, but also the Beef Shorthorn. Ben has always been able to recognise a good backside when he sees one, and sitting beside him during a sale or show is a great eye opener as he shrewdly examines each in turn, in a revealing conformation appraisal, or total write-off. 'That beast's a scraggy devil, no bloody use at all. Lovely head but look at those dreadful legs, and not enough length', he guffaws. 'Half the beasts today can hardly move, I like to see them walking properly, then they will put on flesh.' He could almost be describing the participants of a beauty contest, and not only knows a great deal about good conformation, but also has a memory that holds farm livestock's equivalent of *Who's Who*, as he is familiar with blood lines and pedigrees stretching back into the annals of time.

'I've seen far too many changes and few of them are for the better. *Big is beautiful* is a complete load of rubbish. Our forefathers weren't stupid. These big continental animals need far too much food, unlike the Angus, the Shorthorn and the Highland. Big is bloody expensive. I might have still have had all my nose if it hadn't been so big,' says Ben, who had part of it blown off by a shell during the war.

'In my day, the old Caledonian Road Mart in Perth during the 1950s was the place to be. The whole of Perth came, and during the first week of the sales it was the turn of the Aberdeen-Angus, while the second week was dedicated to the Beef Shorthorn. The old auctioneer, Lovat Fraser, sold an average of one beast a minute. He took bids off the flies on the wall. He was the Prince of Auctioneers. The mart was all cobbles then, and the stockmen slept upstairs with their animals in the straw, or perhaps would doss down in the pens downstairs. They would whoop it up all night, and woke anybody that was trying to doze when they went out at 4.00am to exercise their animals down the streets. Many of the exporters would come to see the beasts as they were walked. That's when they chose their bulls. That cobbled street is now the main road to the A9. In our day it was closed for two weeks because farming

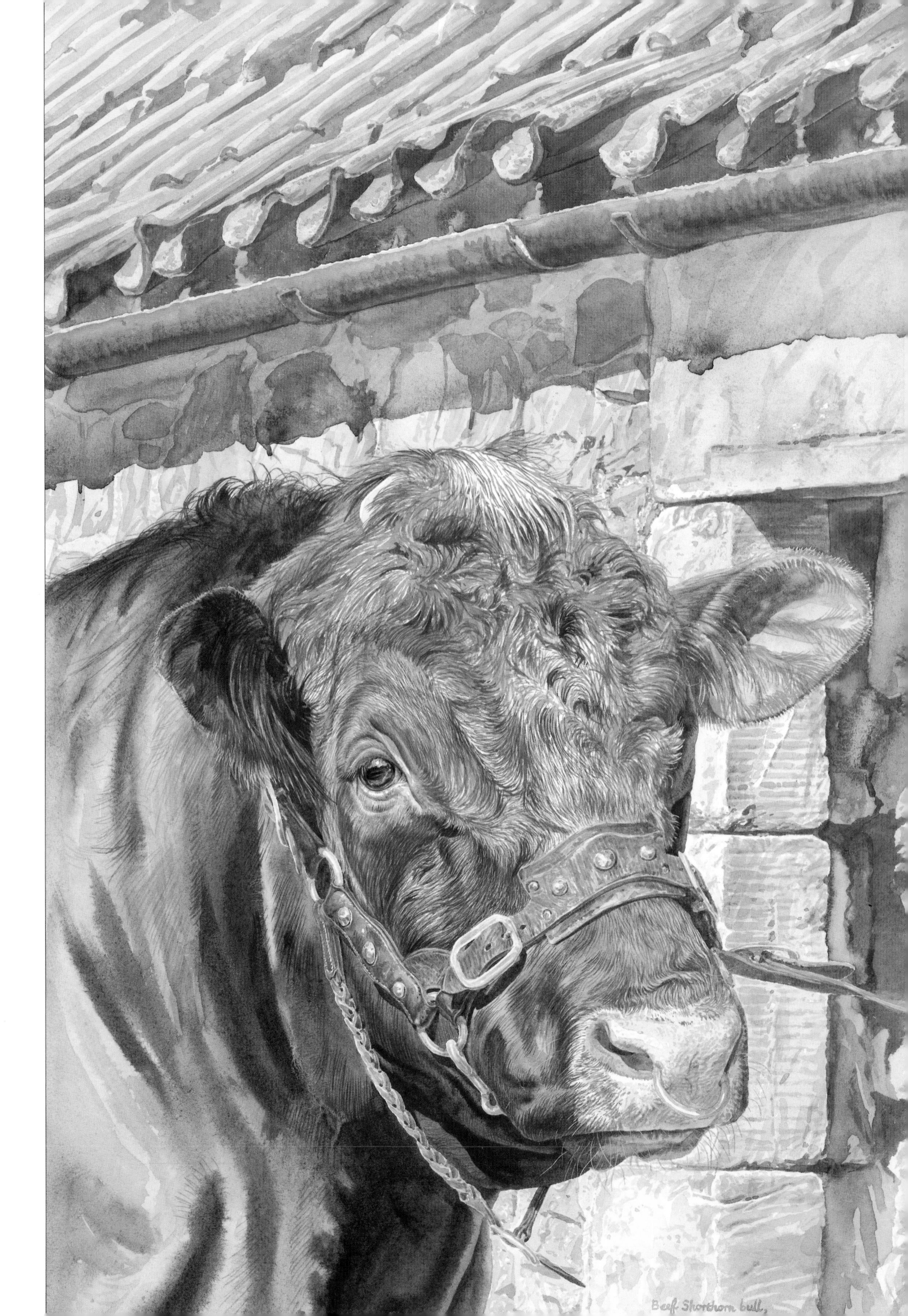
Beef Shorthorn bull,

Cows and calf at Fingask Farm

was so important then. The mart was a wonderful place with lovely stalls, and wooden blocks where the bulls could be tied. Each stall could house four bulls, and it would be padded with hessian sacks stuffed with straw. You used to see the gleeful yardsmen at the end of the sales, leading four bulls apiece. Some of them were as drunk as drums. They always had whisky stored away in their kists. Many of them came down from Speyside, home of all the distilleries.'

While according to Ben Coutts and other traditionalists, big does not always mean beautiful, it is this that the Beef Shorthorn has had to contend with in recent years. Willie McGowan has had to watch the downward spiral that has made the Beef Shorthorn a rare animal. Once nicknamed the *Great Improver*, the breed was still very popular during the 1970s, and in 1974 Willie travelled to Canada for the Toronto Winter Fair, and came back having bought a young bull – Scotsdale Havelock. It was this bull that greatly influenced the McGowan's herd, giving his offspring exactly the size that they required – neither too small as was incorrectly favoured by the Argentinians at that time, nor too large. 'I had to go down to Kirkcudbright to collect Havelock after his flight from Canada to Manchester. That bull had such a great nature. However, despite this he was always funny about microphones, particularly on one occasion at the Highland Show, when he became very upset. Frankly, I always blamed the flight for that.' A simple explanation that conjures up a glorious picture of a young Shorthorn bull sitting in economy class, partaking of the in-flight hospitality, while nervously listening to the air hostess' safety announcement over the tannoy.

(opposite) Fingask Jupiter

Fingask Jupiter

'In 1945 we used to have three men working on the farm and they were paid £27 between them for a fortnight's work, but today all that has changed, and there are few farmhands as it is no longer economically viable to employ anyone. Now I have my family to help, and it is really thanks to them that we continue.' There is such a mountain of paperwork that swamps the farmer today, much of which seems a terrible waste of time; this arduous task is handled by his son James and wife, Margaret. 'Liz is great with the Shorthorns and loves them as much as I do. In fact the whole family has helped. James and my grandson, David, help with the halter training of the show animals, and Liz does much of the showing.' The McGowans have always been involved with showing, not only as competitors, but also judging too. Willie judged at Smithfield Show during the 1980s, and has also judged in Ireland and at many other venues all around Britain. 'Havelock was champion at The Royal Yorkshire Show where we always did well, and it was a sad, sad day when he developed bad rheumatics and we had to put him down. Most of the folk on the farm cleared off.

'The Shorthorn is a hardy breed, and the only reason we bring them in during the winter is purely to stop them poaching up the ground. They should have good length about them, and another important feature, which is inherent in them, is an excellent temperament. All our bulls have rings put in their noses, which makes handling them easier. We used to use "snitchers", which helped us control the bulls, but now rings are fitted by the vet when they are about 18 months of age, and don't seem to put them up nor doon. The Shorthorn should have plenty of bone and a guid smart walk, and for me a beautiful head is very important too. They make a great cross with the Highland.'

While visiting Mrs Jane Nelson who has the Achnacloich Fold of Highland Cattle, we saw one of the Fingask bulls, bought specifically for this purpose. Mrs Nelson had been sad that the Beef Shorthorn was dwindling to such an extent, and told us how good they are to cross with Highlanders. The progeny of this cross have a far higher milk yield, and was also the basis from which the Luing breed of cattle was developed – now a breed in its own right. 'The Shorthorn is a wonderfully versatile animal. Their colours vary, and they may be red, white, or roan. The latter is usually the most popular,' Willie explains, leaning on his stick.

Meanwhile, Liz has brought one of their current bulls, Fingask Jupiter, into the yard. He stands impeccably while the sun warms the buildings, and her Jack Russell terrier charges round the steadings, accompanied by two ageing collies. Despite the antics of the dogs running all round the bull's legs he does not bat an eyelid, demonstrating the superb temperament typical of most Shorthorns. While Liz has now taken over the running of the livestock, including their flock of well-known pedigree Suffolk sheep, James looks after the arable farming. The cropping is made up of oats and barley, some wheat, hay and silage. Fodder beet is also grown for the sheep. Willie, now a widower, has a second married daughter. His oldest son, William, was tragically killed in a car crash.

The McGowan family have ridden the storms that this glorious breed has weathered, and stuck with them through thick and thin. Even though today the Beef Shorthorn is a recognised Rare Breed, there is a current resurgence – hope for the future. Willie McGowan's dedication has paid off; with his daughter Liz following hotly in his footsteps he can rest assured that she will promote the Shorthorn just as well as he has himself all these years.

In all my travels to talk to people about their livestock, no person has left a greater impression on me than James Biggar, Chapelton Farm, Haugh of Urr. I do not think that I have ever met anyone with a greater knowledge of cattle. Together with his son, Donald, they farm 1500 acres of mixed ground in Kirkcudbrightshire. Their pedigree Shorthorn and Galloway herds were renowned all over the world, and up until the year 2001, the Biggar family had entries in the Galloway Cattle Society stud book every year since 1876. For many years they bred pedigree Hereford cattle, and have always been staunch supporters of native breeds. When I asked James what he thought of the Belgian Blue – an obscene looking beast with double muscling that I would describe as the Sumo wrestler of the bovine world, he laughed and said, 'Well, there's one good thing about Belgian Blues, if a bull turned nasty at least you could run away from it.'

When the dreaded Foot and Mouth Disease struck Dumfries and Galloway in the early part of 2001, the Biggar family were terrified of the consequences, for their farms lay surrounded by sheep, many of which had been newly acquired at the market where many of the problems began. But they were powerless to protect their livestock, and one morning James' worst fears were confirmed. By the end of that dreadful, catastrophic day, generations of careful breeding and dedication were turned to ashes. The Biggars lost almost 1000 beautiful cattle. Neither of them dwells on the misery and heartache that surrounded this terrible outcome, and I was humbled by their incredible resolve to pick up the minuscule pieces that were left in order to rebuild their lifetime's work.

'At 90, I suppose I am a bit old to be starting all over again, but perhaps I am just too foolish to do anything else. It will be Donald who will see the results with a new herd. We used to have several different pedigree herds, but it became very complicated. Now we are going to concentrate on the Shorthorns. Donald and I are 100 per cent

Captain Ben Coutts

farmers, and I have had the Shorthorn bug since I left school at 16, when I went out to Canada to help with a consignment of Shorthorns that were being shipped out there'. James is modest about his influences on the breed, but his family has been involved in exporting cattle to Canada, Argentina, South Africa and Australia since the 1890s. 'No single breed can do every job, but the Shorthorn is as good as you will get, and is at the base of most other cattle. It has laid a good foundation for improving other native breeds all over the globe. The Shorthorn has many essential components of an ideal suckler cow as she is docile, a good milker, easily calved, and has excellent fertility. One of the few disadvantages is that when she is crossed with a black bull, the progeny will usually lose its colour, becoming black or grey.'

Mr Biggar's friend, Bertie Marshall, had the biggest herd of Beef Shorthorns in the country. His Cruggleton Herd was famous world-wide, and he exported many pedigree animals every year from his farm in Wigtownshire. The herd was dispersed in 1952, and there was a massive dispersal sale. As a youth, James used to spend much time with Mr Marshall, and clearly learnt a great deal about the export market. Since then he has travelled all over the world, and seen some of the greatest herds of cattle. As well as the McGowans, he mentioned the Durno family of Uppermill, Tarves, Aberdeenshire, as being important breeders of Beef Shorthorns. 'Until we lost all our animals, we three were probably the only old breeders left.'

He clearly remembers the days when the foreign buyers came over to Scotland in January to study the various herds, prior to the Perth Bull Sales in February. 'There were many important buyers, often wearing big hats, and they travelled round Scotland viewing prospective animals. We had a great deal of fun but sadly it has all changed now. At one big lunch party at Cluny Castle in Aberdeenshire, it was a really cold day. There was a log fire, and one of the Canadians got too close to it and burnt his trousers. Later at the Perth Sale he was seen wearing the same ones all burnt up the legs. When he was asked why he hadn't changed, he replied,"it doesn't pay to look too prosperous around here." On another occasion, at a lunch party in Mid Lothian, one of the big exporters, Bob Duncan, was sitting next to the hostess. The talk was entirely about bulls, bulls, bulls – of course. However, she was looking rather bored, so he asked her "aren't you interested in cattle?" to which she replied, "Oh yes, if it wasn't for my money then the Shorthorns wouldn't be here." At the other end of the table her husband quickly replied, "No, my dear, and neither would you."'

The bulls used to come down from the north to the Bull Sales by train with their breeders. On the journey the breeders played cards together in the carriages. However, as the train slowed as it came into Perth Station, and everyone started gathering up their hats and coats, someone said, 'Now gentlemen, this is where the friendship ceases.' There was always real competition at the sales then. 'We won't see those days again,' James laughs.

For someone of 90, James Biggar has an outstanding memory, not only full of fascinating anecdotes, but also filled with breed histories that date back almost a century. He showed me photographs of many of his great animals, and in particular one bull, Tofts Romany, who won the Championship at the Royal Highland Show two years running, and also won the Royal Show. He and nine of his sons were lost in the cull. All that the Biggars have left is some stored semen that they will use in the future when they start to rebuild their herd. 'Do you think your artist could do something with one of these photographs?' he asks me as we study the magnificent bull. It is a poignant moment that summarises all the misery and sadness – yet James is full of hope for the future.

After the catastrophic outcome of Foot and Mouth Disease, Donald and James quickly decided that they must waste no time in rebuilding their Shorthorn herd. Donald, also an eminent livestock judge, chairman of Smithfield Show, and vice chairman of Quality Meat Scotland, knew of many good Shorthorn bloodlines in Canada, and managed to acquire the nucleus for a new herd. Then in a brilliant, forward thinking plan, he arranged for frozen embryos to be collected and sent back to Scotland. These were implanted into cows at his sister and brother-in law's farm in Perthshire. Obviously, it was many months before the Foot and Mouth restrictions were lifted, and animals could be re-established in Kirkcudbrightshire. 'We had not collected any semen from our bull, Tofts Romany, until he was nine years old, but took some from him at this time intending to lay it down like fine wine for use in perhaps 10–20 years. If there was only one bull whose semen we wanted, this was it. It all seems so ironic as we had no idea then that it would be all we would have left. It is now gold dust, and we will eventually use it. We also managed to obtain some embryos from the mother of our stock bull, Special. He was Romany's successor and had come from Canada. We are hoping to have between 100–120 calves born from the embryos we have implanted. If there is a silver lining, and I emphasise the word *if*, it is that we will have imported a great deal of good new blood into Scottish Shorthorns, and this was badly needed,' Donald explained to me. Once again, I saw the incredible resolve and dedication that makes James and Donald Biggar such brilliant livestock farmers. It is a fascinating story that illustrates how modern scientific methods can retrieve an otherwise impossible situation in the wake of an earth-shattering disease. No two farmers deserve more success.

When the *Scottish Farmer*, a farming newspaper, came to photograph James Biggar for their Living Legends feature, he stood in front of the farmhouse with some Shorthorn cows behind him, and was asked what his greatest achievement had been. His reply summed up the total humility of a man who has probably done more to promote and support the Beef Shorthorn than any one else. He never mentioned his OBE, nor the massive long tally of championships and accolades that his animals have won, nor all his numerous personal achievements, nor his important work in improving and maintaining many of Scotland's native cattle. He said his greatest achievement was staying alive, and sticking to Shorthorns when every one else was getting out of them. I feel greatly honoured to have met him.

The end of an era, Tofts Romany at Chapelton farm, Castle Douglas

CHAPTER SIX

The Crofter's Pony

Mary McGillivray is a kindred spirit. We first met many years ago while we were moving house. I had four tiny orphan hedgehoglets to hand rear and was desperate to find some goat's milk for them. Mary kept goats; however, when I spoke to her she was very unenthusiastic about giving me any milk, until I casually mentioned that it was not in fact for human consumption, but for baby hedgehogs. 'Oh! If it's for hedgehogs you can have as much as you need, any time.' Our subsequent meetings left me in no doubt that Mary and her husband Donald are two rather special people.

In 1978 the McGillivrays had their first encounter with the Eriskay pony, and they have been working with dedication to save this extremely rare animal ever since. 'We first noticed them in the Rare Breeds Survival Trust's tent at the Highland Show, and thought that it would be a real tragedy if this hardy island breed should disappear. When we met them we realised straight away that they had such nice characters, very much part of the family,' Mary explains.

During the latter part of the 1980s, the Rare Breeds Survival Trust introduced Category 7 for a short period of time. This was a watching brief for breeds that did not fully qualify for inclusion in the Priority List, but were under consideration by the Trust. Although there was considerable interest in the Eriskay Pony, it was still waiting in the wings and had to fulfil specific criteria before being recognised as a native breed.

It was on the remote Outer Hebridean Island of Eriskay, which lies between Barra and South Uist, that the last few small 'Western Isles' type ponies were discovered. Eriskay's inaccessibility had safeguarded these remnant animals from the influence of different bloodlines, whereas many ponies from the other islands had been cross-bred over the years to produce a larger animal which could pull heavier implements and carry more weight. However, with such extremes of climate, it was soon discovered that the original wiry little ponies were better suited to weathering the storms. Transporting ponies from island to island was no mean feat, although occasionally a few were trussed up together and closely packed on an open craft. Transport difficulties probably helped to save this ancient pony breed from extinction and were one of the reasons that a few survived on Eriskay.

Dr Robert Hill, who was the GP on the island of Barra for over 20 years, fondly remembers the arrival of three Eriskay fillies. He had told a fellow doctor friend on South Uist that he was looking for some ponies, and shortly afterwards took delivery of three ponies, described as being totally different from the widespread garrons. 'That was in 1964, and we were very ignorant then about the Eriskay pony, but it was obvious that these animals were indeed different. I remember they had such wonderful temperaments, and we could do anything with them. One day we wanted to carry stobs and wire out for doing some fencing on the croft. We had some pack saddles and put them on the ponies, and then gently loaded them up. They went like a dream,' explained Dr Hill. 'Now, dear, your memory seems to be slipping,' interjects Mrs Hill. 'Well you know what I mean, considering the ponies had never done that sort of thing, they were really very good.' The Hills reckon that it was these three ponies that raised awareness to the fact that the Eriskay was a breed in its own right.

A dedicated group of people set out to save the last 20 ponies and to establish them as a recognised breed, a vital part of Scotland's heritage. Following a great deal of work and research this has happily now been achieved, and the Eriskay was officially accepted as a breed in its own right in 1997. For the Eriskay this was a dramatic breakthrough, and it is now classified by the Rare Breeds Survival Trust on their 'critical' list – Category 1. It is the wish of the Eriskay people that a nucleus of ponies should always remain on the island, ensuring that the animals remain true to type.

'They were originally crofter's ponies and carried peat and seaweed. Someone was also reputed to use his pony for carrying gas cylinders. Usually handled by women and children, the ponies went out on the common grazing from the first of May to the first of November, and lived at the doors of the crofts during the winter months. This is why they are often referred to as "back door ponies." Eriskays are very people-friendly and human orientated which makes them so special. The island people are very keen for us to find a market for them, unless we do, no one will breed them. They need a redefined role as a family pony, because they do make wonderful riding and driving ponies, easily ridden by children, being less broad than some other native breeds. Recently a couple from Cheshire, Lesley Cox and Andrew Gooden have had national success driving their Eriskays as a tandem,' Mary explains.

In 1979, a newspaper article stated that the MacLachlainn family from South Uist were paid an allowance of 75p per day to fuel their daughter Seonag's school transport. There was no road to the croft and six-year old Seonag travelled to school on a very environmentally friendly school bus, an Eriskay pony. The only mounted Scout troop used Eriskay ponies on the Isle of Uist, where they once again proved their adaptability.

The Eriskay stands at between 12 hh. and 13.2 hh. and should not be too broad. It has light boned legs and less feather at the heel than the more substantial Highland pony. They tend to have a fairly large head unlike the small dished Arab-influenced head of breeds such as the Welsh pony, and their jaw is big enough to contend with rough herbage. Colours are predominantly grey, although blacks and bays are occasionally seen. Foals will often be born black or brown but will eventually turn grey as they mature. Eriskays are athletic little animals and are extremely sure-footed and strong, and well able to carry a small adult. They have extremely hard feet and an active gait. Though their coats are not long, they are dense in the winter which protects them from the high rainfall of the Western Isles.

In 1979 Mary and Donald acquired their first Eriskay pony. 'We were asked if we would like to take a pony, with the explanation that it would also be coming with its daughter which was half Eriskay and half Highland. Of course we accepted, because they were so near the brink then. A friend went to Oban

Braincroft Sundew

to collect the ponies in his trailer. They had travelled by boat from Benbecula in an open topped container and were all plastered in spray and salt from the sea. What we did not know was that the daughter also had a foal at foot, a Shetland cross. The night that they arrived, we bedded them down in the byre, and they looked very relieved and happy after their stressful journey.

'Over the summer months, Ciurstaidh, the daughter, became fatter and fatter and we kept on cutting down her grazing, until one day she produced a second foal, Magnus. Clearly there had been a Shetland stallion running amok on Benbecula and he had obviously found himself a hill,' laughs Mary. 'Seamus, the first foal, was very naughty and lay down when the children were trying to ride him, but we found Magnus much nicer.'

Peigi, the McGillivray's original mare, was a great character and went on to live until she was in her mid thirties. 'We decided we would like to drive her. Donald and his brother made a flat breaking-cart out of a Morris Minor axle with some Oregon pine for the shafts. While I was away, they thought they would like to try it out, without any thought to the preparation or to breaking the pony to this new task gradually. However, they were very lucky, and, in true Eriskay fashion, Peigi took to it without any fuss and was soon driving us to the village or to check the sheep. Later I was asked to give rides in the cart to children at local events. By this time the cart had a proper seat made from government surplus pine, where you perched up high and felt like the Queen. One day I was busy giving rides at Abbeyfield, in Comrie, and Peigi was behaving impeccably when suddenly the pipe band arrived. I did not know much about ponies and pipe bands at that point, having been brought up in England. When she heard the first skirls of the pipes as they started to tune up, Peigi started to dance about and was suddenly on her toes so that I felt it would be much safer to leave. We went over the Ross Bridge in Comrie like a rocket, at a spanking trot, and straight on to the main road without stopping until we were within sight of home. Everybody adored Peigi as she was so good with children. Once I found her surrounded by them, with one little boy between her back legs and the cart, but she would never have dreamed of kicking anyone. She gave rides to dozens of children who came to visit the farm and was adored by them all.'

Peigi had many foals by a stallion the McGillivrays were lent called Ballachan. One of her sons, Braincroft Fingal, has proved to be quite a lad. At 18 months of age he went to Kirkintilloch, where he served many mares before going on to Eriskay, where he spread his genes round the island. From Eriskay he travelled to Kent, and from there to Cornwall. After this, the Genus Stallion Centre at Whitchurch took semen from him and other Eriskay stallions to hold in safe keeping. Mary was given a very special rosette bearing the inscription: 'Braincroft Fingal – For Services Rendered Nationwide.' This well-travelled pony is now serving mares in Banffshire. His activities have clearly been of paramount importance in saving the Eriskay.

Braincroft Sundew, 'Sunny' for short, was five years old when we saw her. She was a pretty grey roan colour that will turn paler in time. Mary is enjoying riding her and has recently acquired some traditional creels, once used for carrying seaweed on Eriskay. These sit across the pony's back on top of coir matting, and are held in place with rope and a round, smooth stick like a broom handle which sits beneath the pony's tail. While the nearest Sunny may be to seaweed is the supplement that Mary feeds all her ponies, she is perfectly at home wearing the panniers that her island ancestors once wore to carry seaweed from the shore. Traditionally seaweed was used all over the islands as a mineral rich fertiliser full of trace elements that provided an ideal balance to the sandy soils close to the sea.

In his book *Scotland's Native Horse*, Robert Beck has described Donald McGillivray as 'an enthusiastic breeder.' Mary and Donald have two sons. While this statement may perhaps conjure up the wrong picture, there is no doubt that the McGillivrays' passionate dedication to this ancient pony that can be traced back to Pictish times, has played a vital role in saving them from an uncertain future.

We travelled to Eriskay on the ferry from Ludag on the southern tip of South Uist. A grey seal

Grey seal

poked its head lazily from the icy waters as the ferry crossed the Sound of Eriskay, which has since been spanned by a causeway. This now means that the islanders are no longer dependent on the weather, and boat timetables, and can easily come and go as they please. The ferry man was most obliging when we told him of our interest in the ponies, and borrowed a car from the pier to race us up the hill to see his own three ponies. Slim and wiry, these hardy beasts were grazing the short turfy grass, but trotted willingly down to see us. His old flea-bitten grey mare was heavily in foal, her belly distended. Well into her twenties, this was probably to be her last.

We were then deposited on the doorstep of the manse. Despite our visit being unannounced, Father Callum McLennan, a native of the island, welcomed us with typical West Highland charm. 'Those of us who remember the Eriskay pony as a working animal will be eternally grateful to them for their contribution to crofting life. When I was a boy here, there were Eriskay ponies all over the island. Now there are merely about two dozen. They are very special ponies, so friendly and good with children. At the moment they are far out, on the township's grazings.'

A clock ticked rhythmically, chiming the half-hours with an electronic bell that seemed oddly incongruous in this timeless location. 'Did it make you jump?' Father Callum enquired as I glanced round to see where the noise was coming from. 'Every one wonders about my clock,' he added with a mischievous twinkle. Outside, perched upon a ladder, a pair of legs chatted away in Gaelic to someone else up on Father Callum's roof. 'All the church services here are in Gaelic, it's our first language on the island. We don't have many incomers; however, you can tell where they live because they have flowers in their gardens. When I was a boy, the garden was for growing food. Why would we need to plant flowers when the Good Lord has provided us with such a profusion of beauty in every ditch and bog?' he laughed, pointing to the yellow flag iris beds and bog cotton drifting cloud-like in the salty breeze.

He gave us a whistle-stop tour of the little island, pointing out the local landmarks and the famous beach, where once Bonnie Prince Charlie had landed. The salt water had obviously been penetrating the engine, and the car back-fired, spluttering and struggling up the braes as Father Callum put his foot flat to the floor. 'Nearly everyone here comes to church, but forty per cent of the islanders now are pensioners. It's a wonderful place, half the time we don't even bother to change the clocks. You should see it on a good day when the bays are turquoise and the sand is white; it mirrors the colours of the heavens.' Father Callum left us with that thought, as we walked lazily back to the ferry, enjoying Eriskay's peace. As his car disappeared over the horizon in a cloud of fumes, I couldn't help wondering if that was what was meant by 'Holy Smoke.'

Despite the fact that the unique Eriskay pony has many staunch supporters; in the year 2002 there were still under 100 registered breeding mares. The future of these small island ponies is far from secure, and without doubt they still require as much help as possible.

PICTISH PONY

Spring in rural Perthshire, with white clouds in the sky,
Blossom fills the air with scents, and the larks are flying high.
Fields are striped with patterns, made by seed drills, ploughs and harrows.
Mallard on the river bank nest where the water narrows.
She stands beside a stable now, but once there was a croft,
Her forbears lived an island life, and carried peats aloft
With panniers made of willows, and filled with fish and stores,
The island's precious ponies spent their days beside the doors.
But in the name of progress, new ponies came to stay,
They tried to breed them bigger, to make them pay their way.
But soon the new arrivals had pushed them to the side,
And Eriskay's precious ponies began ebbing with the tide.
And in the name of progress, the tractor came as well,
As all the old traditions, collapsed, then died and fell.

This island in the Atlantic, swept by gales and surf,
Had houses thatched with marram grass, or built with roofs of turf.
The men went off to sea to fish, the women worked the land,
While Eriskay's children rode to school, or played upon the sand.
An island often storm-bound and cut off by the weather,
Had ponies round about its doors, no need for rope nor tether.
This ancient Pictish pony was suited to its life,
But alas its numbers dwindled as it witnessed years of strife.
And the crofters have long gone now, and the thatch has blown away,
In its place the bungalow, broods dully every day.

While this quietly dozing pony with panniers at her sides,
May never carry seaweed, nor listen to the tides,
Her home's a grassy pasture with trees to stand below,
Her transport is a horse-box, while the crofters had to row,
But little does she realise she still carries a great weight,
Pictish pony, Eriskay, must be saved from uncertain fate.
She may never work the croft again, nor graze the salty land,
But her precious foals will safeguard the hoof prints on the sand.

CHAPTER SEVEN

The Long Sheep of the Hills

THE Scottish sheep breed that suffered the worst losses during the horrifying outbreak of Foot and Mouth Disease that swept across Britain during the autumn of 2000, and on into 2001, was undoubtedly the South Country Cheviot. This extremely hardy little beast is thought to be not only the oldest recognised sheep breed in the country, but is also famed for its extremely wily nature, excellent carcass quality, and beautiful wool.

However, today's South Country Cheviot has been much altered and little resembles its Celtic forbears. With influences from many different breeds, including the important wool producing Merino, this alert-looking, white-faced sheep with small erect ears, has a long frame on short, stocky legs, and is perfectly suited to the hardships of the Cheviot Hills that divide Scotland from England. So perfectly is she suited to this landscape etched with hill burns, small glens, scree slopes, heathery moorlands, and rocky outcrops, that her loss from the area would be grave, for hundreds of years of careful improvement and history would be lost for ever.

When officials came knocking at the door of Tim Douglas' farm, Upper Ashtrees, at Jedburgh to make arrangements for the culling of his entire Cheviot flock, he was struck with terror. Like the Cheviot ewe, there have been Douglas' in the area since time immemorial, and with a passion not only for his sheep flock, but also for all things rural, Tim was not prepared to go down without a fight. After making a few urgent telephone calls, he quickly discovered that not only had the MAFF officials made an error about his flock and come to the wrong farm, but that they had also made their gruesome decision based on the location of one tiny meadow pasture that was within the contiguous culling area. This offending piece of the Cheviot landscape had not had any livestock on it for months.

In an amazing set of circumstances, the Douglas' proved that their flock was not affected in any

(above) *Tim Douglas*

way, and with the help of a sensible District Veterinary Officer, they miraculously cut through the miles of red tape synonymous with the whole Foot and Mouth saga, and saved more than three hundred years of ovine history. Devastatingly, many others in the area were less fortunate.

Tim Douglas is an exceptional person. My initial telephone conversation with him to arrange a meeting left me in little doubt that he was most certainly the person I wanted to see about the South Country Cheviot. So it was that I set forth to Jedburgh on a December day when the whole country-side was painted with glorious glistening hoar-frost. The frozen water in my windscreen washers, and low winter sun made driving while peering through the diminishing peep hole on the windscreen well-nigh impossible. Despite this hazard, the landscape was looking at its best, dressed like a sparkling winter bride, high-lit by yellow shafts of sun that twinkled on every branch and twig. A party of rooks busily engaged in probing a steaming roadside dung heap for breakfast seemed to have changed their usually drab plumage for shot silk, as if attired for a Hunt Ball. The views everywhere were simply stunning.

On arrival at the farm, I was greeted by two amicable Border collie bitches, and then received an equally effusive greeting from their owners. While Tim's charming wife Jane plied me with sustenance, he and I immediately fell into a fascinating conversation about rural issues as a wonderfully memorable day unfolded. Once we had chewed over strong feelings about the government's appalling mishandling of livestock health issues, and the spine chilling story of his flock's brush with the funeral pyre, I soon realised that I was dealing with a man of great passion, someone who would fight to the death for what he believed in, but with a fairness of mind and good humour that I had seldom encountered.

Tim began pulling out papers, magazines, photographs, poems, records and catalogues – enough material in fact for me to write an entire book about his involvement with the Cheviot sheep. He showed me a photograph of a particularly special ewe at a show. 'That reminds me of a story about Tom Elliot, a well-known figure in Cheviot circles. He was watching a class of ewe lambs being judged at Kelso Show, but the judge seemed to be taking forever to make up his mind. After what seemed an eternity, Tom said, "If this bugger doesn't get a move on, they will have turned into gimmers." It was very funny.'

Tenants of the Roxburghe Estate for over 300 years, the Douglas' hill farms lie several miles away from their home patch at Upper Ashtrees, and we therefore drove away into the far hills to see the rest of their farming enterprises at Mainside and Greenhill. These amount to 3000 acres. Here their flock is 'hefted' to each area. Hefting is an important concept that evolved as a result of flocks grazing vast tracts of unfenced lands. Following intensive shepherding to retain each separate flock in its rightful area, the sheep were trained to remain on a particular piece of hill. As their offspring were kept and maintained using the same methods, they too have evolved to stay within a given area, thus continuing the tradition. Even today the sheep must be properly shepherded in order to make them use the land to their best advantage. It was these traditional hefted flocks that gave such cause for concern during the FMD outbreak, as their loss would mean that this inherited instinct would be gone forever. Not something understood or valued by suit-clad office officials who probably don't even own a pair of wellies. Little did they seem to realise that this is indeed a vital part of our heritage.

Unless straying off the soulless main roads that allow traffic to pass through the Borders at breakneck pace, the casual passer-by may never be aware of the huge tract of hill ground that forms the backbone between Scotland and England – a glorious piece of country punctuated with tiny little glens, and riddled with ancient Iron-age hill forts, burial sites, and old stone stells – the local term for a sheep fank. Steeped in history, this Border Country formed the backdrop for the activities of the infamous Border reivers who pillaged their way across the country stealing livestock while embarking on skirmishes that have undoubtably shaped the area.

Jedburgh, and many other Border towns, was also famed for its woollen mills and weaving industry where Cheviot wool was once in high demand. Now, sadly, the town's industries are in sharp decline. In the past most farm wool cheques would have paid the shepherd's wages. Now wool, like so many other valuable, useful commodities, is worth very little. I stopped briefly at one of the local woollen mills, and cringed as I found it bedecked in '*Hey you Jimmy bunnets*', badly made tartan dolls, blow-up Highland cattle, knitwear full of nylon, and garish plastic jackets, that only a gullible foreign tourist would contemplate. An accordion version of Jingle Bells was bellowing out of the sound system while the accompanying picture on a large video screen showed a kilted, bearded hero in a Braveheart shirt playing whilst crooning on the edge of a very familiar loch-side. I didn't, however, see any monsters. While I could buy a sticker with the slogan *I love Bonny Scotland*, emblazoned on it, a pure Cheviot wool jersey was out of the question. I beat a hasty retreat.

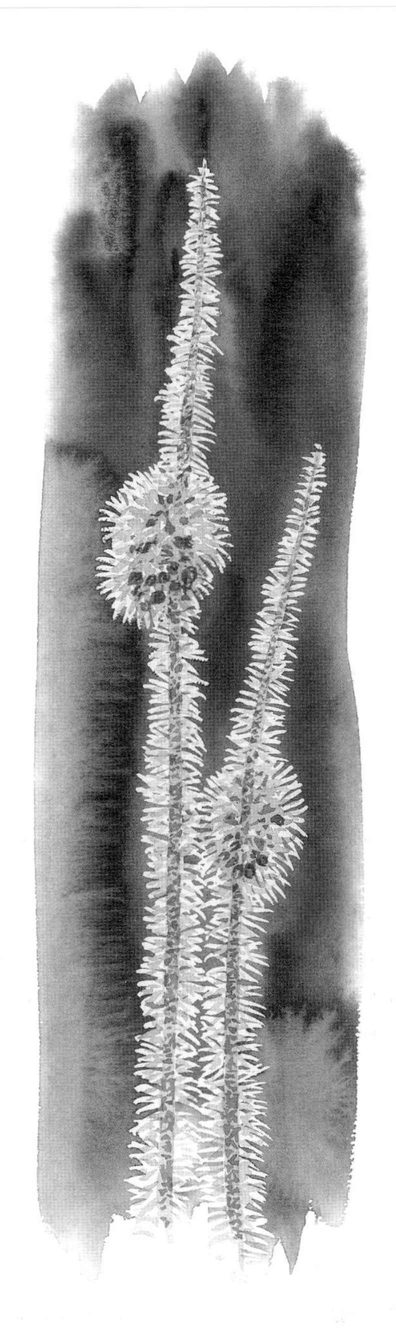

As Tim and I continued to wind our way through the narrow hill road, squeaks and yelps came from the back of the Land Rover indicating the mounting excitement of the two collies. 'Without the Border collie it would be impossible to shepherd Cheviots; they are vital to all we do here, and I admire them greatly. I have to admit that I quite like it when they misbehave too.' Tim, pipe constantly in mouth, pointed out the myriad landmarks, and many of the attractive little farms that were casualties of the dread disease. The road was fringed with crisply frosted rushes, and the bracken glowed bronze in the low light. Cobwebs on the fences were iced with cold breath. Suddenly we spied a fox, back-lit in the afternoon sunshine, coat gleaming deep red gold. Tim stopped the vehicle, and momentarily the animal stopped too to have a good look at us before loping off at an unperturbed trot down through a farm gateway. I have always admired and loved the fox more than any other British mammal. No animal could prove so resilient. And unlike some of its scabby town dwelling relatives, this one looked in the peak of good health.

Soon many of the crossbred sheep we passed were replaced by the pure bred Cheviot – distinctive, alert and impressive in its native surroundings. By now the stories were flowing; bubbling forth like champagne at a society wedding. Tim seemed in his element revealing a mine of local knowledge. While his father's family have been in the area for generations, his mother was a Shetland crofter, and his father, now in his 90s, used to send sheep up to the islands. Today Cheviots are still numerous in the Shetlands, although they are often crossed with other breeds. Tim and his wife Jane have four children, clever and successful, but none of them farmers. 'Sad though this may seem, I am in many ways relieved, as the whole system of agriculture has gone to pigs and whistles, and who on earth can say what will happen next.' An extra large puff of smoke drifted out of the Land Rover window as Tim accelerated on up the hills.

'There is nothing wilder than the Cheviot ewe. When we have to lamb a gimmer, it is always better to tie her leg to the heather, or a stone, or else she will bugger off. But once she has licked her lambs she will be alright. Otherwise she will head straight for the burn with her tail flying round in a circle. People think that the Blackie is the wilder and hardier. While she is certainly more prolific, in really bad conditions the Blackie will plummet. In 1947, 1963, and 2001, we had some really bad

Ewes and lambs with stells, Heatherhope

snowstorms, and hundreds of ewes were buried in the drifts. In these situations the Cheviot will come into her own. She may not have the most massive lambing percentage, but records show that she is always consistent. Years ago the Cheviot was known as the *Long sheep of the hills*, while the Blackie was referred to as the *Short sheep of the forest*.'

We passed a large dam where herons stood sentinel on the hillside, and a pair of goosanders drifted on the water. 'They bring bee hives up here during the summer months. Do you know there is a large herd of feral goats up here too, and during the Foot and Mouth epidemic though they were culling farm animals everywhere, the goats were wandering wild and free, but no one did anything about them. When a vet came round the hills checking the situation, he said we should walk round our march-fence. He was over here after working in the Sudan and was not in his first flush of youth. When I told him that there was over 3,000 acres of ground much of it at over 1,800 feet high, he quickly changed his tune.'

We crossed on to a hill track beside an old railway carriage with a series of tumbledown pens around it. 'That's Belsen: as my uncle's shepherd was helping out just after the war when he saw the state of a very thin gimmer that had been left in a pen and forgotten, he remarked "it's fair bloody Belsen up here", so the title has stuck. I don't know how far we'll be able to get up the hill as they were shooting up here the other day and have cut the ground up dreadfully, so we'll probably stick.' At this Tim took to the quagmire with a particularly determined expression which I felt had stood him in good stead. The vehicle bounced and ground her way up the steep slope in true Land Rover style, while Tim puffed, and pushed her into diff lock, leaving the little glen far below.

On the hill-top, a massive panorama unfolded like several sheets of Ordnance Survey maps stretching far beneath and all around us. The collies leapt out with glee and quickly discovered a lone walker perched against the fence, quietly contemplating the world while eating his sandwiches in thickly gloved hands. After a few territorial barks, the dogs realised that he was good for some last crumbs and a morsel of ham. But our unannounced arrival had clearly interrupted the musings of a loner on the rooftop of the Cheviots on a day when they were looking at their best.

Tim knows every hill and valley, every landmark and horizon, and as the sun fell down to the west, and a sharp breeze ruffled our jackets, he bombarded me with names and dates. In a flowing of Border terms, laws and rigs, knowes and cairns, shanks and hopes, were all unveiled before me. Here were words that painted pictures even without the scene at my feet: Muckle Sundhope, Beefstand, Carter Fell, Hungry Law, Peel Fell, Dere Street and Pineilheuch. Names with a past – places and hills that perfectly suit the wiles of the Cheviot sheep.

In the distance we watched as a peregrine falcon was mobbed by ravens as it sped to a roosting place, and against the opaque sky, glided, and then continued in a fast clockwork flight. 'There are lots of adders up here, buzzards and blaeberries, and badgers too have increased greatly in recent years. Often the badgers end up being road casualties. That hillside there is a favoured place for ring ouzels, but the lapwings and curlews have diminished, and the last blackcock were seen a few years ago now.'

Tim employs two shepherds who we met on our descent. Craig Weir, a great Cheviot man with shepherding in his blood has been at Mainside for well over 20 years. He proudly showed me some of the hogs in the fank. He had already picked out his show hopefuls for the following spring and explained the attributes of each of them. The

Cheviot is a late maturing animal, and years ago was not put in-lamb until she was three years old. *Southies*, as they are nicknamed, may occasionally have horns, an acceptable throwback to past ancestors. Occasionally too, a coloured lamb may appear, this is also a legitimate breed variation. But the instantly recognisable trademark was the alertness of carriage, and ever-listening erect little ears, and the great length of the body. These vital animals standing against the stones of the fank would have immediately caught any judge's eye. New transportation legislation and quarantine periods will make showing very difficult, and sadly stop many farmers from competing altogether. Craig, knowledgeable and dedicated to his work, clearly has a passion for showing, and is hopeful that things will change.

When we returned to the farm the following spring, the lambing was just over, but a vixen had caused bedlam killing many lambs. 'We just couldn't find the culprit, and had to sit up for many long, cold nights. Then one morning one of my collies came round the hill with a lamb's leg in her mouth, and I retraced her steps and found the den. But it was very depressing to lose so many good lambs before we managed to sort the fox out.' Craig explains. 'Cheviots are my life; the Blackie is just horns and hair. However, I must admit I do go off them for about four weeks during the lambing as they can be so thrawn. It is their bright alertness and their conformation that I love. And they are so wonderfully cocky. It is sad to think that nowadays farming seems to be more about numbers than about quality, and that really worries me. Scotland has always had a reputation for quality, and I don't think we should ever let that slip. I am really concerned that people are not being encouraged to farm for stockmanship's sake, and that is a real tragedy.'

During our visit to Mainside, Craig clipped one of his Highland Show competitors. 'We clip them as early as we can so that they will have enough wool on them for showing. I always use hand shears, as they don't leave the skin so bare, and generally do a far better job than electric clippers. Some of the show gimmers are clipped in April but then have to be kept inside for a while. Though I love the Highland Show, it is our local show, Pennymuir, that I would rather win than anything. There is always such great crack with the lads there. To me that's what showing is all about', Craig explains.

There is a great sense of community in hill areas, and it is therefore a real tragedy that people are being driven from the countryside due to the hardships of earning an honest crust. This is a sad twist of fate that oddly mirrors the disastrous affects of the Highland Clearances. Then it was the arrival of the Cheviot ewe that drove the people from the land. Towards the end of the eighteenth century, many Cheviots were taken up to the far north of Scotland – to Sutherland, Ross-shire and Caithness. Gradually these sheep were also altered and evolved to suit the different ground of the North Country. This gave rise to a separate breed of Cheviot, a larger beast, the North Country Cheviot, fondly nicknamed the *Northie*. With the rural economy in crisis, and the loss now of many remote communities and their people, it seems ironic that today's issues greatly pose a threat to the existence of all our native farm animals, for without people there can be no livestock. 'Mules have been the ruination of the sheep industry', Tim laughs at this statement, 'but there seems to be a new trend that hopefully will mean eventually it will be quality that counts and not quantity. Then the Cheviot ewe will once more come into her own.' Tim is optimistic.

As we travelled back down towards Upper Ashtrees, we passed a vast area of excavation. A bleak reminder of the horrors of Foot and Mouth Disease, for it was here that thousands of animals were burnt in pyres that smoked for months, filling the atmosphere with acrid, choking smoke. 'People were paid £80 a ton for their straw to keep the fires going. Ironic that it was the best paying thing that farmers had had for years.' Huge bags of ashes awaited safe disposal. But where? They were all that remained of treasured flocks and herds – people's livelihoods. Bloodlines discarded like kitchen waste. To me they symbolised the foolishness and error of man's greed.

Of all the people I have met in recent years, Tim Douglas' devotion to rural life is probably more staunch than any other. A Justice of the Peace, he is also a wonderfully talented poet as his rhymes on

country matters show. In 1990, the Cheviot Sheep Society celebrated its centenary. Appropriately he wrote the following poem. It perfectly summarises all that surrounds this wonderful old breed, and more than shows his skill as a poet, his humour, and perhaps above all his total understanding of the ovine mind.

Craig Weir

THE WEE WHITE SHEEP

Tonight we unite with out comfort assured
A love-hate relationship, sealed and endured:
A century gone and another ahead,
To honour the creature our forefathers bred.

They sent her up north when the Highlands were cleared.
It was her, more than redcoats, that Highlanders feared.
She took to the north as the clansman departed
And the Cheviot finished what Cumberland started.
They altered her genes to suit their land's need
And she mothered the North Country Cheviot breed.
Back home on the Border her value stood high
As they crossed her with Leicesters for stocking in-bye.
So the lowlands were stocked and the industry grew
With prosperity based on the Cheviot ewe.
As wild as her namesakes, the Cheviot Hill,
She has turned us demented, near broken our will.
Her instinct maternal, when startled, new-lambed
Is near non-existent, and thus she is damned
To every expletive that Borderers learn
As she runs like a greyhound and heads for the burn.
We have thrown dogs at sticks and attempted to fell her
As her tail went rotating most like a propeller.

We've clipped her and cursed her and nursed her when ill.
We've dipped her and lambed her and damned her to hell.
We've bought her and sold her and caught her and told her
That beauty is aye in the eye of the beholder
And that if she transgresses there's aye the deep freeze
And she looks at her best with mint sauce and green peas.

Yet, let us give credit where credit is due
And drink to the health of the Cheviot ewe.
Though her lambing percentage may not be excessive
Her lamb's conformation is truly impressive.
In times of recession, her quality tells
For her lamb is the lamb that most readily sells.
For the century past she has been paying our bills
And ensuring that Borderers stay in the hills.
Let us pray she continues, our heather to stir,
For our way of living depends upon her.
So charge up your glasses, stand up and drink deep
To the next hundred years and the Cheviot sheep.

Tim Douglas 1990.

CHAPTER EIGHT

A Trek with the Highland Pony

THE Highland pony is one of the most versatile and hardy of native breeds, and I always thoroughly enjoy my meetings with many of the characters associated with them. Though I did show Highlands while working with them, how preferable it is to accompany them in the field in which they truly excel: riding long distances up over the high hills carrying a picnic and a camera. For me the Highland reigns supreme as a riding pony having the ability to cross the roughest terrain. There is something very special about being out miles from anywhere, enjoying the scenery while being totally at one with an animal. But I shall never forget my acute embarrassment at Perth Show when my mount bucked and farted her way round the main ring as we were being judged, leaving me in an undignified heap round her neck. Then worse still, when the infamous judge climbed aboard, she galloped out of the ring with a flaying of hooves in a cloud of dust, totally out of control. I stood in the middle of the ring chewing the end of my riding crop as casually as possible while listening to a tirade of Anglo Saxon oaths coming from my departing beast's jockey, then sprinted out in search of my steed, and a very disgruntled judge. My own Highland mare, Bridie of Alvie, was a great joy. Kind enough to carry the smallest child, she was a much-loved family member.

The Highland pony also makes an excellent carriage-driver. During my early twenties I often acted as groom for a friend who drove competitively. We had more than a few hair-raising moments as we tanked off through obstacles while competing in marathons. The driver herself was pretty fearless and we used to nip along with mane and tail flying in the wind, the competitive spirit very much to the fore. During a big competition at Scone Palace, we were going through some of the obstacles on the side of a steep brae face. I am fairly small, but the driver at that stage of her life, was a very large lady who has since become half the woman she was. We were flying down the hill, when suddenly the pony

(above) Cameron Ormiston

slipped over to the wrong side, losing her footing in the churned-up ground. I was perched on the topside, with the cart listing at a very precarious angle. 'Lean, lean, lean you bugger quickly,' shouted my driver who was anticipating disaster. A spectator on the hazard told me afterwards that everyone was doubled up with laughter at the sight of little and large, with little hanging out from the top-side of the cart endeavouring to stop it toppling. Apparently I looked as useless as a chocolate fireguard, and my feeble efforts to counter-balance the cart were in vain until suddenly the skipper lurched herself over to my side of the cart, momentarily squashing me beneath her as she narrowly averted a tragedy. The pony meanwhile had held her own, and soon we were off again at a spanking trot. But my mouth was dry with the adrenaline of the moment.

A lack of spirit should never be associated with the Highland pony. While hunting in Renfrewshire many years ago with friend Audrey Barron, we had a truly memorable day. Audrey was riding a vast heavyweight cob called Smokey-Joe, while I had a wonderful little Highland mare, Aileen of Dryden. We had shampooed the two hairy beasts thoroughly the previous day and loaded them white, pristine, and gleaming. When we arrived at the meet, they had managed to cover themselves in green stains during the journey and looked pretty messy. A bucket of cold water and vigorous application of a dandy brush did little to remove the marks. While we were trying to tidy them up, they danced about on the spot, stood on our toes, barged and shoved at us rudely, and generally in their excitement made life very difficult as they pawed the ground and thoroughly misbehaved. Audrey had not been out with the hunt before, and found the whole inauguration very daunting. Everyone else was mounted on perfectly turned-out, impeccably mannered thoroughbreds, clipped and stain-free. We stuck out like two sore thumbs, especially as our mount's behaviour was drawing even more attention to us, as they pranced about with flaring nostrils and heads tucked in like two Trojan War horses. Furthermore, the whole atmosphere of the hunt was having a very laxative effect on the pair of them. When a rather smart lady appeared with a tray of sherry and sausage rolls, I politely took the sherry while Aileen grabbed a sausage-roll, neatly plucking it off the tray and nearly knocking the remainder from her hand. Smokey-Joe meanwhile was doing something not dissimilar to the Highland Fling, dancing about on the spot because he had been briefly separated from Aileen.

Sally Coutts with Chester of Alvie

When we finally set off with a couple of blasts on the master's horn, Smokey-Joe shot down the road with a large buck and accompanying gut sound effects, as he overtook the field, and worse still the master. Poor Audrey hung round his huge neck in a heap while a disdainful voice informed her that it was a cardinal sin to overtake the master. Sadly she had no option, for Smokey-Joe had found unstoppable strength. As the day wore on, a chapter of disasters left many of the thoroughbreds lame or exhausted. One horse stopped at a wall while his boozed jockey sailed over and had to be carted off to hospital. Meanwhile our two were still going on like two steam trains, and whinnied frantically at one another each time they were momentarily separated. We crossed a field of Shetland ponies where hunt jumps had been specially built into the fences. Here an immaculately dressed woman was battling to coax her big horse over a relatively small jump, and we had to wait while she tried and tried. Waiting with two thoroughly overwound horses was like holding on to the fuse on the end of a hand-grenade. Finally as she was just about to admit defeat, six woolly black Shetlands came galloping up the field in a neat line, barged past her and popped adeptly over the jump. After this our two flew over too and nearly unseated us in their enthusiasm to follow the Shetlanders. We galloped on up a track totally out of control. Massive clods of mud flew up in my face, thrown up by the dinner plate size hooves of Smokey-Joe in front. Suddenly we saw a cattle grid ahead. My heart was in my mouth. We sailed over it like two rockets. I still shudder to think of that moment. Though we certainly would have had no marks for keeping up with the Jones' that day, our two supposedly *unsuitable* mounts were still in at the finish, and could have galloped on for a good deal longer. Anyone who thinks the Highland pony is a boring trekking animal, is sadly misinformed.

The Highland pony is still widely sought as a garron, and used for carrying the deer off the hills during the stalking season. In Glenartney, Perthshire, as on many other Scottish estates, the ponies still work in a traditional manner. All these working ponies are shod with special shoes with heels that give them extra grip while they are carrying heavy weights, and stop them from slipping on rough terrain.

It was in 1960 that Alistair Work, now the head stalker, came up as pony boy following a visit to his house from the head stalker of the time. 'Willy Bennie was a wee man with a Hitler-style moustache, and a Willie Woodbine permanently burning his lips. He appeared at our house and said "I'm looking for a pony man", and I had no hesitation in accepting the job. We get some pretty bad snow in the glen, and half the time I think he didn't believe I would get to work. But I had a Douglas trials motorbike at the time which was great in winter, and even managed quite deep snow. There were 30 ponies up the glen then and they were wonderful beasts. Previously, there had been 150 when the estate bred all its own stalking ponies. They used to use a big black Dales stallion on some of the mares. The Dales is another great work breed. I remember my first day out with three ponies tied head to tail. Then there was no road into Strath-na-glen at all and there was thick mist right down to the valley floor. It was early October and the stags were roaring all around us, but we couldn't see anything. There was an army major out for the day, and Willy told me to wait while he took him off to find a stag. They went into the woods where there was more roaring echoing about. Of course we didn't have a radio or anything, so I just had to hang around and wait till I heard a shot. Suddenly Willy started waving a hanky, and there was this enormous 17 stone stag lying in the woods; that was my first stag. Sometimes you could wait nearly all day, and I remember once Willy kindled a fire about half a mile away as a sign that I had to take the pony over to pick up the stag.

'Willy always hated getting up in the mornings but was great at night, and sometimes we could be out on the hill well into the dark, but oh Jesus what would bring him home was running out of Willie Woodbines! He would fly into the kitchen and start raking about in the cupboards like the devil, getting more and more crabbit, while his wife would innocently ask him what he was looking for. The relief once he found his cigarettes was amazing.

'The rats used to be terrible up at Auchinner when I first came here. There was one particularly

Alistair Work, Ben Vorlich, Glenartney

Bridie of Alvie

cheeky one that used to jump up under the sink in Willy's kitchen to get out. Then it would pop back, but leave its tail showing. One day whilst he was washing a cup at the sink, he saw the tail, "Right, I'll bloody well sort you," he said. He quickly snatched a pair of pliers and grabbed its tail, and then with the rat dangling, he stabbed it with a toasting fork. The rats used to keep you awake at night running about, but in the end the pest control came in and cleared them, and we have never had a problem again.'

Breaking a pony in for carrying deer can take a great deal of time and patience. The ponies that have been broken to ride are usually easier than the ones that have not been previously handled. A deer saddle is very heavy and it helps if the ponies are not too fat as otherwise the saddle will slip with the deer ending up round the pony's tummy. 'Some of the modern day Highlands end up as broad as can be, and frankly it can be like putting a saddle on to an orange. If the saddle slips then it is no use at all and I've known us get into some right messes with overweight ponies. The trouble is they are such good-doers and will get fat very easily, even on the poorest grazing,' Alistair explains. 'It doesn't matter even if the pony is quite high, as long as it stands quietly we can usually load the stag. In Glenartney we lie the beast on its left-hand side, and then bring the pony in from the right. Two of us lift the stag's head on to the saddle and then slide the body round. It will then be strapped on round the pony's tail, and also have a forestrap at the front. The average weight of the stags here is between 10–14 stone, although at the start of the rut, they can be a good deal heavier, and some of the stags that have been living in the forestry become very large indeed,' Alistair explains.

In recent years, deer numbers have risen hugely, and it is now necessary to cull higher numbers. Without this, over-grazing becomes a serious problem, and many animals will die of starvation and other health problems associated with over-stocking. During a hard snowy winter, it can be difficult for the stalkers to cull enough deer, as access to the hills becomes almost impossible. The work for the ponies is very tiring, but despite this most of them seem to thrive on it.

In Glenartney the ponies were once used for every task. They carried panniers to bring the grouse, rabbits, hares, and other game home in the days when there were large organised shoots. They were also the main means of transport in and out of many glens in Scotland – an integral part of life in remote places.

But ponies, like vehicles, can play up. The Glenartney ponies are still frequently used to carry guests out on to the hill as well as carrying the deer home. On one particular occasion, Alistair remembers a guest with a shooting stick that he insisted on taking with him. It dangled at the back of the pony, getting in the way. As they were passing over a particularly rough piece of ground, the rider inadvertently gave his mount a tremendous crack on the backside with it. 'Well, that pony would have won the

Derby. It took off with the old man on top bouncing around and clinging on for grim death. All we could do was watch helplessly. Finally they disappeared into a deep bracken bed, and first the pony emerged, then we found the man sitting stunned on the ground. Another pony had panniers on and as he was coming home through the gate, he got a fright and caught one of the panniers, and it sheared right off. And the fat ponies that don't hold a saddle properly, well they have given us lots of trouble over the years. I remember one pony with a stag on and it had sweated heavily so that the girth had come loose. Suddenly, just as it was going down into the burn to cross, the whole thing slipped and the poor pony ended up with the stag underneath it. Mares are not nearly so useful to us as the geldings, because when they are in season they can disrupt everything with squealing and kicking and all that carry-on. We had a string of ponies out one day, and the one in the middle was a mare. The burn was in spate and she just would not go through. Finally after lots of messing about, old Willy Bennie stuck some thistles below her tail. I can remember him wading into the water in desperation, leading the string through with water right up to his bollocks.'

A pony man may take some time to learn his trade, and though much of the day may be spent waiting, being good at handling the ponies can make all the difference. Sometimes high winds get up and frighten young ponies, and when the mist comes down it can be very hard to find the way home. Loading the stag in an effortless way also takes practice. After more than 40 years, Alistair knows this only too well. Fitness is important, and it can be as bitterly cold as Siberia. 'One year a very stocky, broad-shouldered rabbit trapper came up to help us as a pony man. Now Willy Bennie was just a wee fella, and he was explaining how to put the stag on board and said, "put a good bit of effort into it." Well the rabbit trapper certainly followed his instructions and put a huge amount of effort into it, and the stag just flew right over the top of the pony and straight on to flatten poor Willy on the other side. I wasn't there at the time, but Willy recounted the tale and luckily thought it was very funny.'

Alastair and his wife Janet are well-suited to glen life, but have without doubt seen many changes. He remembers clearly when there were large blackcock leks with sometimes as many as 40–50 birds at a time, and numerous capercaillies too. With changing habitat there are few blackcock in the glen today, and capers are extremely rare. Both the Works are very devoted to the ponies, many of which go on to a ripe old age and stay out on the hill living a semi-wild existence well into their dotage. Often at this stage a new home will be found for them on lower, softer ground, where there is abundant grass and where they can be housed in winter if necessary until the end of their days. For much of the year the Glenartney ponies live in a large herd with great freedom. During the winter they are well-fed and are always given hard feeding while they are working, but in the spring and early summer, their days are spent dozing lazily on the carpets of wildflowers, or paddling in the hill burns. These are ponies that have a good life.

If you asked Cameron Ormiston of Newtonmore, what were the qualities he looks for in a good Highland mare, he would give you that mischievous look of his, and tell you that he likes his ponies exactly as he likes his women. They must have a pretty head, a kind eye, alert intelligent ears, fine attractive hair, a nice clean leg, and attractive feet. They should be in proportion and must certainly not be overweight. Last, but not least, they must have good action. Cameron has a reputation for being keen on the ladies, so he should know.

A well-known figure in Highland pony circles, Cameron and his forbears have spent their lives in the hills and glens of Scotland, and know the ways of deer, ponies and grouse. Cameron's grandfather, Edward Ormiston, was the head-stalker at Gaick at the head of Glen Tromie, Invernesshire, where he also bred Blackface tups. The family were always associated with good Highland ponies, which were not only used for carting the red deer off the high hills, but also as a main form of transport to and from the estate.

All the Ormistons' ponies had to work hard for their living. They were often transported miles in a flat cart before starting a full day on the hill. Edward bred a fine black mare called 'Mountain Polly'. In the summer of 1911 he walked her 14 miles to Kingussie station, and then took her on the train to Glasgow. The Royal Highland Show was at Paisley that year, and she won the Overall Championship before setting off back on her long haul to Gaick to continue with her daily work.

Cameron's father, Ewan, sometimes had to walk the 18 miles to school in the village of Insh. The family's first language was Gaelic, and little or no English was spoken. Ewan was the dog boy for his father before becoming under-stalker and then going to the First World War, where he was twice wounded and decorated in the field. On his return to Gaick, he applied for a keeper's job at Aberfeldy. Having struggled to make his way to the interview after miles of walking and a lengthy train ride, he was turned down, simply because he was considered too young for the job. He swore to his parents that he would never work for anyone in his life, and never did.

He opened a butcher's shop in Kingussie and supplied much of the district, including many of the travellers who passed through. All the meat was of local origin. Venison was popular at the time and was often exported to Germany. Local grouse would be shot, plucked, dressed, and packed in boxes of white heather to be sent to New York soon after August 12th, ready to appear on the dinner table at the Twenty One Club. He never liked to see anyone going hungry, and eventually had so many bad debts owing to him that he went bankrupt in the 1930s.

During the Second World War, Cameron and his father took the stalking on many of the big estates all over Scotland. With so many men away fighting there was a great need for people to cull the deer and take out the stalking parties. The Ormistons travelled across Scotland to places as remote as Knoydart, Islay, Rannoch and Achnasheen. Their ponies were in great demand, and each year they rented out between thirty and forty for the duration of the stalking season. One of the heaviest recorded Highland red deer stags of the time came from Glenfiddich and weighed 25 stone 4 pounds. Much has changed on Highland estates; Glen Feshie, a neighbouring estate, had 18 ghillies and 22 ponies at that time, and it was said that one of the biggest costs was the bill for the ghillies' whisky. A ghillie's wage was 30 bob (£1.50p) a week, but this was supplemented with perks of hares, venison, salmon, and housing.

During the Second World War Cameron's parents moved to Glen Banchor Lodge where they took in evacuees. The evacuees had to be inspected for lice, and Cameron used to cry if he did not have them, for despite the tickles, having lice ensured time off school. Cameron has always been infamous. When he was born, he was reputed to be one of the heaviest babies in the area at 13 lbs. Indeed, it was the village doctor who gave him his first pony, a black Shetland by the name of Betsy. 'She took off with me on board one day and dumped me outside the hotel door with my kilt round my head!'

Highland ponies had more than proved their worth on their native rough terrain, and many were required for the war effort. More than 800 ponies were rounded up from all over the Highlands and Islands. They were invaluable and were used by the Scottish Horse Regiment and the Lovat Scouts for carrying heavy loads of equipment during both the First and Second World Wars. The Highland pony has always been a truly versatile animal. 'Highlands are very bright and they don't forget easily,' said Cameron, who spent many long evenings breaking in the ponies. 'Our breaking methods were rough and ready, and some of the ponies could be quite wild. Some of the worst ones came from Otter Ferry when their owner was forced to give up. There were thirty of them and we went with three lorries. It took five shepherds and twenty dogs to round them up into the old buildings. They were as wild as hawks. On our way home we were caught in a terrible snowstorm and were forced to spend the night in the lorries. It was quite a night.

'When I first started working with my father, I was paid the same as the ghillies – thirty bob. A pound of this went to pay my mother for digs, which left me with the equivalent of fifty pence. I supplemented my income by poaching, and selling the game back to my father for his shop. No one

ever asked any questions,' Cameron grinned wryly. After the end of the war the family took over the Balavil Arms in Newtonmore.

Soon they had set up a successful trekking operation, with customers coming up from the South of England for an action packed holiday. 'In 1952 the grand sum of twelve guineas could provide you with a week's pony trekking across some of the finest scenery in Scotland, with cooked meals, packed lunches, tea and cakes before bed, and your boots beautifully polished. One week we had a group of bonny nurses. They had all just received their exam results, and had passed. Well, we rode up to one of the bothies and were cooking them some food, when they started bringing the drink out of their rucksacks. It was a hot day and they kept on drinking. It wasn't long before everyone was starkers in the river.

'Some of the trekkers were unbelievable riders. I can tell you I've seen them all ways: hanging off low branches, snagged up in the brambles, and some really thought they knew how to ride. One particular lady came all dolled up in smart brown leather boots, spurs, and all the gear. And she was very bumptious. We used to allocate each rider a pony for the week. We had one infamous mare called Loch Eil, who liked to lie down in the water. So the lady with the smart brown boots was given Loch Eil, and right in the middle of the river Spey, down she went. That sorted it,' said Cameron with a hearty laugh. Some summer evenings Cameron and his friends used to lure the girls up to the local graveyard to hear the echo from the surrounding hills. 'It could be quite spooky up there, and then they would hold your hand,' he added, squeezing mine.

The trekking season was from April to August, and then the ponies were used for stalking. The Ormistons had 200 ponies during their heyday. Cameron also took people out skiing in the winter, and clearly enjoyed teaching some of the ladies. 'Anything that was good, well, you never took it in small doses did you?' he laughed again.

The Ormistons' ponies have always been a commercial venture. They were some of the first Highland ponies to be exported. From Leith docks ponies went in large crates on to the deck of a ship, and then on to wagons on the train, to travel to Germany. 'We travelled with them, at the other end of the wagon. We have sent a great many ponies abroad. When you are in it for commercial reasons, every pony has its price.'

Cameron's own homebred ponies have excelled in many fields. In 1966, his stallion, Glen Tromie Trooper, went to the Horse of the Year Show at Wembley where he was Reserve Champion, with his proud owner cutting a dash in full Highland dress. 'I always wanted to win the Royal Highland Show with a black pony, and in 1989 Ebony Polly of Croila was Female Champion, although she did not win the Overall Championship. We used to show much more, but now I've cut the number of ponies right back and am down to fifty.' An icy north wind was bringing squally showers as Cameron took us round endless fields of ponies. His favourite colours are yellow and mouse dun, though he has Highland ponies of every colour. They were standing hard up against the grey stone dykes, sheltering from the blast, with their long manes and tails blowing in the wind, but trotted up to see us, careful not to miss the opportunity of Cameron's bucket of feed. 'I usually have plenty of sweeties in the car too for all the local kids, only one of my sons has told me that I'd better be careful nowadays. It's a real shame, because I love children. I once had a group of them out from Glasgow. They came by bus. We were up the hill near a wood and I said, if anyone wants to spend a penny they'd better go now, and one little voice piped up and said, "please sir, where's the shop?" I love kids.'

Cameron has judged Highland ponies at the Royal Highland Show, the Royal Show at Stoneleigh, and the Paris Show. He also found himself called upon to judge horses out in Russia, much to his amusement. He is always full of devilment; at a meeting of the Highland Cattle Society, when everyone was worrying about the dangers of the dreaded disease, BSE, Cameron was heard to suggest, 'Why don't we start eating Highland ponies, they've never had the disease.' Of course, this was said in jest. Unfortunately, not everyone thought so.

Cameron's sons are following close in his footsteps. Dochie has taken the Gaick prefix for his own ponies, while Rhuraidh's ponies have Craig Dubh as their prefix. All his family work with ponies, sheep, or Highland cattle. Recently Cameron had a major worry with his health. After his test results came back, he said to the doctor, 'I want to know how long I have left, because I have a lot of damage to do yet.' Happily, he has made a full recovery and is still doing damage in style.

CHAPTER NINE

Hebridean Sojourn

THE livestock mart is a great place for meeting people, and it was there in Oban that we bumped into Mrs Ena McNeill from North Uist. As the crowded market thronged with bustling activity, and the hot, sweet breath of the Highland cattle filled the pens, we were introduced. A most attractive lady, with a round and smiling face like a Cox's orange pippin, a special aura surrounds her, and is clearly evident from the first meeting. When she invited us to visit her in Uist, we jumped at the chance.

And so the following summer we prepared Keith's ageing camper van with everything we would need for our trip. We drove to Uig in the far north of Skye, from where we took the boat to Lochmaddy, on North Uist. En route we enjoyed several stops. While eating our breakfast on a birch-fringed loch, a pair of black-throated divers drifted out of the mist, and it was only the voracious appetite of the ubiquitous midge that put an end to our bird-watching. After numerous photograph stops, we realised that time was running short. Skye is a surprisingly long island with dramatic scenery and jagged mountain ridges, and the photographer is beckoned insistently. We consequently had to race the last few miles to Uig, and just caught the ferry by the skin of our teeth.

The first night we found a spit of land that ran out into the Atlantic. It was a perfect resting-place to which we returned on several occasions. On one side a large brackish pool was a great source of intrigue. Three mute swans fought an enduring battle as one was continuously ousted from the ménage à trois. Many other birds came and went, splashing, preening, feeding and squabbling, or merely just drifting on the pool's ever-changing surface. A black-tailed godwit paid a fleeting visit, while on the rocks by the shore, a party of whimbrel probed for food. Glorious oyster-catchers, pied and red, whistled past the windows as we ate our supper, and in the bay before us huge parties of

Ena McNeill

Lapwing and chicks with machair flowers

shelduck and eider bobbed on the tide with rafts of ducklings. One night an otter ran across the sand. And finally as the wildlife dispersed for the night, the play of reflected light on the water coloured our dreams with magical, rose-gold hues.

Another evening we parked on a windswept headland high in the dunes. As the rice was simmering on the gas stove, and we were supping a glass of wine, a school of dolphins emerged in the bay. Their lithe shapes swirled up through the water in acrobatic play, while the sun, a grapefruit orb, fell into the Atlantic Ocean, and painted the sky with a pattern of cream, gold and yellow, sending rays of light in bronzed ripples across the water. We watched hypnotised until the last chinks of light vanished in the west leaving us with only the haunting cry of a curlew echoing over the dunes. At dawn we were awoken by the sweet trills of the lovely, rounded corn buntings, perched atop lichen-covered posts. Uist. A touch of heaven.

If it were not for the inclemency of the climate, the Hebrides would probably now look similar to the Costa del Sol. So after all we can be grateful for the vagaries of our appalling weather. Eternal beaches remain deserted; livestock and wildlife footprints are often all you will find. And there is not a deck chair in sight. Far from modern civilisation, unspoilt by commercialism and cheap tourist lures, if you can put up with the weather, the Hebrides will provide you with unique, soul-restoring memories.

As the wind played in my hair, while I watched a lapwing brooding her chicks framed by purple orchids, plantains and buttercups, she was serenaded overhead by her iridescent mate, wheeling in a sky of racing clouds. Golden globes of marsh marigolds lined the burn as redshank darted back and forth in frenetic activity. During the summer the flora of the Hebrides is breathtaking as the machair bursts forth in a series of colour phases that begin with yellow. The *machair* is a rare and precious habitat only found on the north-western seaboard of Scotland and Ireland. Formed over thousands of years by gales blowing shell sand in off the Atlantic Ocean, it provides a unique environment for a wealth of plants and animals. *Machair* is a Gaelic word that means a low-lying fertile plain; a lime-rich habitat that lies behind the dunes, it depends on a careful grazing system. Livestock are only put on the machair during the winter months. The Uists have some of the best examples in the British Isles. Its flower-studded grasslands are dramatic, and lure thousands of breeding shorebirds each year, including dunlin, snipe, redshank, curlew, oyster-catcher, lapwing and ringed plover.

Ena McNeill's love of the island's flowers is clearly apparent, and she is very aware of the sensitive management that is so necessary in order to provide this ideal habitat for a profusion of flora and fauna. Carpets of orchids, many of which are unique to the Hebrides, and frequently hybridise, are sometimes hard to identify. Others take on a slightly different appearance, often battered by the rigours of the weather. Marsh orchids and the early purple orchid grow in abundance, jewels with emerald green speckled leaves.

Ena has spent all her days living and working on their traditionally run croft in Kyles, North Uist. While her family were poor, her crofting father was adept at providing them with all the sustenance they needed. 'Father was a great poacher, and despite our lack of money, we always ate like kings. I used to poach too, and learnt a great deal from him while taking ducks, cormorants, rabbits, and flounders. Hunger was something we never knew, until the moment I was sent away to school in Inverness, which I hated. The matron was prosecuted for taking our food, and I loathed the hostel where we stayed as I was so homesick.' Ena has a mischievous twinkle that lights up her whole face, as she reminisces about her childhood.

She has her back to us as she chats while milking a lovely black Highland cow. 'When I was 11 and 12 years old, I used to look after the cattle to ensure they did not stray as they were never fenced in, and I would sit by myself with the collie and play my mouth organ. I have always adored animals, and longed to be a vet. By the time I was five, I could milk a cow. At home, we have only ever drunk the milk from our own Highland cows. It is so delicious, and it has a high butterfat content.' The milk swishes into the jug as the cow contentedly munches her way through a bucket of oats. 'This cow came from the Isle of Mull, and is called Smeorach of Callaich,' Ena explains as she snuggles her head into the cow's long woolly black flank, and the jug fills frothily. 'I used to dream of having a few lovely Highland cattle on the croft, and now my son Angus MacDonald and his wife Michelle and I have over 200.' Her twinkle at this statement reminds me of a child who has just had all her dreams fulfilled. 'Without Angus, I couldn't have survived. While he was still a lad, he worked harder than any other on the island, and that's not boasting.' Like his mother, he loves animals, but is also passionate about farm machinery, and according to Ena when he was just a teenager made one good tractor out of two old wrecks. 'Angus loves cultivating too, and helps many people all over the island.'

The family farm their land using traditional methods, fertilising it with seaweed that they have collected from the shore. The seaweed – tangle – is dragged up the beach, usually in the late autumn when the fierce gales that lash over the islands wash large quantities ashore. It is then taken in a tractor and trailer high above the tide-line where it will be left to rot. 'I used to do it with the horse and cart. We spread it in February and March, and it can go straight on fresh for corn, but for tatties it needs to be well rotted,' she explains. They are both keen to see the revival of the seaweed industry. During the 1970s, they would have been paid £120 per dry ton of tangle by the factory at Barcaldine where the weed was used to produce gelatine, cosmetics and food. The factory also provided much-needed jobs for the islanders.

The islands have a manganese deficiency. Normal oat seed will not grow, and so the islanders must keep their own seed, usually black and sandy oats, to plant the following year. Rye and barley are also grown, although the cattle are not so keen on the former. Rye is a useful crop however, as it will grow on scorched or bare ground, and can be mixed with oats to make it more palatable. All the harvesting is carried out with the binder and thrashing mill.

Ena has over 80 hens that roam free, scratching on the shore. As we chat inside now, and Keith starts making some sketches of her, she is enjoying a freshly beaten egg from a glass. 'It's very good for you,' she explains. She is renowned for her culinary prowess, and was visited by the well-known chef, Nick Nairn, when the BBC made a programme on the islands. Angus had to pretend to catch the flounders that they had caught the previous day, as the weather was so unreliable. She makes crowdie – a cream cheese made of soured skimmed milk, and dumplings using her mother's old recipes. Her black and white puddings have only the most natural ingredients, and are reputed to be richly succulent. Certainly Nick Nairn was suitably impressed.

But it is their glorious fold of Highland Cattle that play the most important part of Ena's life. Each animal has a name, and most have a history accompanied by a wonderful story. It was in 1975 that they started the pedigree Ardbhan Fold. Having some spare time before catching the ferry from Oban back to Uist, Angus and Ena had gone into the market. 'Angus wanted me to bid for a lovely 1000 guinea heifer. However I said, "if I sell the car I might afford it." But the thought of riding back to Uist on a Highlander's back did not appeal to him, so we bought Fluran Og of Glen Nevis, a cheaper, brindled heifer which we paid for with money from 10-year-old Angus' piggy bank.' From then on they continued to buy two or three animals every year, funded from the proceeds of Ena's thriving bed and breakfast business.

Now they have Highland cattle of every different colour and type: black, red, brindle, white, and dun – colours that blend perfectly with the windswept moorland landscape. They provide a glorious spectacle as they cool off in a turquoise sea on a warm day, while overhead a vast sky is filled with wispy white cloud. We accompanied Ena as she went to feed two young calves. As she called across the waving beds of yellow flag iris, they came bucking and skipping across the bog cotton, and quickly drank their bottles of milk. A corncrake uttered its persistent rasping call, hidden amid marsh marigolds by the burn, while a redshank on a nearby post called to its mate. A large black cow stood against a huge grey rock surrounded by flag iris, cudding peacefully while the salty breeze rippled her long coat in waves.

Ena and Angus think nothing of rising through the night to check on their beloved cows at calving time. Selling them though, is hard, and the animals have to travel by ferry to the mainland, and then to the market at Oban. Though the Highland Cattle sales are one of the highlights of the year, parting with them is not something Ena relishes. In 1988, she sold her first pedigree bull, Pibroch Dubh. 'I had a call the night before the sale, with an offer of £1500. I was very torn, for I really wanted to go to the sale. After I had turned it down, I panicked that I had made a big mistake. However, the boat journey, on the *Lord of the Isles*, went well, and when we got to the sale, I changed into my kilt and put tartan ribbons in my hair, and in Pibroch's too. Well, the bidding started at 1000 guineas, but the auctioneer was very good at promoting this small, naturally reared black bull from Uist. Suddenly the bidding went through the roof, and my heart nearly stopped when it reached £2500, and the auctioneer said, "and he's going back to the island." The bull lifted his head up, and in tears, I kissed his nose. So back he came, and the crew of the *Lord of the Isles* joked because he'd been on the boat before.

'Pibroch then spent four years on Benbecula before going down to England. It was wonderful that someone from the island bought him. We then bought one of his daughters, Lily Anne, for my grandson's fourth birthday. It was extraordinary as not only was she by my bull, but she was also out of a Corrymuckloch cow, a blood-line that I had always wanted.'

Many of Ena's favourite cows have ended their days on Uist. Lassair, an old, arthritic beast who had damaged her hip by falling into a rabbit hole, would wander down to the sea every day to paddle up to her belly. Though she was very crippled latterly, she was still clearly enjoying a good quality of life. 'My cousin called her the three-wheeler. I always used to keep a lookout to ensure she was all right as

Highland cow, North Uist

sometimes she found it hard to get up after she had been lying down. One day, my neighbour came round to tell us that poor Lassair was drowning in the sea. I ran out to fetch Angus, and we grabbed a rope as the tide was coming in fast. I told him that he was not to risk anything by going out into the sea after her. After all at the end of the day, she was just a cow, and I knew she had to go one day soon. Suddenly Angus started to laugh telling me that I was a stupid idiot. "It's the neighbour's big black labrador having a good swim after one of his rabbiting excursions," said Angus. We both laughed when we saw Lassair standing on the beach further away. She was a very wise old cow and lived until we had to put her down at 22. I had to go away that particular day.'

Other members of the Ardbhan Fold are well-travelled. Originally from the Isle of Mull, six black Highlanders were exported to Canada in 1988. However, due to health rules and regulations they returned, eventually ending up on the nearby Island of Barra. Ena heard about them and bought four, finally acquiring the last two as well. 'They are lovely cows, and we always keep them all together. I felt sorry for them after all that travelling.' This is typical of Ena who always thinks of the welfare of the animals.

Angus has recently rented Vallay – a huge tidal island about 10 miles north of Bay Head. This 600 acre area of machair provides perfect wintering ground for about 130 cattle. 'The fields are very beautiful there and had not been ploughed for over 40 years, but the soil is very rich and he has planted corn and potatoes. In the spring the primroses are a sight to be seen, and I love the bright yellow of bird's foot trefoil that flowers later in the summer too. We have to take the cattle off by the first of May, and they make a spectacular sight crossing from the island at low tide.

'Crofting has changed a great deal in recent years, but we believe that animals should be farmed as naturally as possible. You can put Highland cattle where you would not put a heavier breed, without having to pump them full of concentrates. They are designed to fend for themselves, though we check ours very frequently. It will be good when people realise that the traditional native breeds are really the best,' Ena explains.

After another afternoon spent in Ena's company we retired to the van, now parked on the dunes overlooking Kirkabost Island. Keith picked up a sketch he had made hastily as we drove over the moor in the morning. We had stopped as we spied a short-eared owl perched on a pedestal of cut peat. It sat unblinkingly while he deftly captured it through his telescope, attached to the van's window. Short-eared owls are frequently seen hunting by day, but it had remained motionless in all its glory, yellow eyes shining as the light broke through the clouds. A pair of hen harriers quartered the ground below, while the eerie cry of the curlew carried from the shore on the westerly breeze. As he picked up his sketch and scanned it with a critical eye, a corn bunting started to sing. Once a common farmland bird, it has now become a rare sight on the mainland. However the Hebrides remain a stronghold. Lapwings waltz together over the carpet of clover, silver weed, buttercups, and daisies,

Short-eared owl

spot-lit by the low rays of the sun. It looks as though the evening will bring another soul-searing sunset. A whisper of a breeze teases the marram grass on the dunes, as the waves on the shore erase our footprints.

Letterboxes in the Uists are not only used for mail. The large flocks of starlings that congregate all over the islands find that they make the most comfortable sheltered nurseries in which to raise their broods. Being showered with letters does not seem to be a sufficient deterrent; letterboxes here have flaps over their openings to keep the birds out. Telephone boxes, too, are frequently occupied by starlings. Bespattered with avian graffiti, and with doors occasionally ripped off by the gales, island vandalism is usually the result of natural causes.

As we drew up at Balranald, the RSPB's Reserve on North Uist, a corncrake immediately began making its repetitive rasping call. This unmistakable sound drifted across the blooming machair amid calls of redshank, oyster-catcher, and that most evocative bird sound of all – a drumming snipe. With the door of the van wide open, it was a good moment to put the kettle on before Keith began to paint the machair flowers, and I went for a wander round the glorious headland. Two more vehicles drew up, and as they parked, and their occupants opened their doors, the corncrake began again with precision timing. In a flurry of activity the new arrivals opened their car boots, and hauled out impressive-looking camera gear. Laden with the weight of their lenses, they began to stalk towards the persistent rasps. The calls stopped. Nothing. Two more cars appeared. Unbelievably, the same scene was played out again. Keith looked up from his sketching and said, 'There's a pressure pad on the road that sets off the recorded calls of the corncrake each time a new car appears'. We giggled at this silly suggestion as once more the new birders grabbed their camera gear. Oddly, it seemed even larger and more imposing than that of the others now standing scanning the machair with their binoculars. The corncrakes were silent. We laughed again as we watched the charade, and suddenly a bird began to rasp again. This time the sound was quite far away, coming from the old graveyard on the hill. In hot pursuit, and with no hope of even so much as a photograph of a corncrake's wingtip, the birders trooped off. Like intrepid explorers thirstily looking for an oasis in the desert, they rushed away in the direction of the last calls, the weight of their equipment ensuring that they would probably need to visit an osteopath once they returned home. Photographing corncrakes is no easy task. The secretive birds had no intentions of revealing themselves. But like a carrot on a stick, their calls continued to tantalise and torment the poor tourists.

I walked round Balranald amid carpets of daisies, sea campion, thrift, and pale yellow heartsease pansies. Plantains waltzed in the wind. The sea was so blue, and yet so green. And the sand so white. Alive with bird life, the sky was filled with white clouds like a flock of round sheep chasing one another through an azure field. The salty breeze ruffled my clothing, and I could feel the freckles breaking out on my face. Hougarry, a small village, sits at the mouth of Balranald Bay, and from this magnificent viewpoint, the Ferguson family's house perches near the shore.

Donald Ferguson is a graphic artist with the Education Department, but his many hobbies include a passion for traditional Scottish breeds of livestock and poultry. He has small flocks of Soay and Hebridean sheep, and has been greatly responsible for promoting the Hebridean both at home and further south. Thanks to his tireless work, the Hebridean has now been re-established in its native homeland. In his garden coops of fowls were well-pegged down to withstand the battering of the gales. Contented hens and a cockerel or two strutted about on the seaweed on the sand below the garden fence, ensuring that their eggs would have the flavour of the shore, with yolks as gold as marsh marigolds.

Donald breeds Scots Dumpies and Scots Greys. The Dumpy is, as its name suggests, a short-legged fowl with ancestry that dates back as far as 1678. It almost became extinct, but in the 1960s was saved from doom by the importation of some stock from abroad. Dumpies are usually black or cuckoo coloured, and have found popularity once more, being not only attractive but fairly good layers. Short-legged fowls have been recorded in Scotland for over 200 years, and had a variety of nicknames including, *crawlers*, *creepers* and *bakies*.

A Scots Grey hen was sitting tight on her eggs in an old sink filled with flowers at the back door. The plants had long wilted. It was obvious that the breeding business was far more important than floral decoration. Here was a Hebridean Kitchen Sink Drama waiting to unfold. Taller and more elegant than the Dumpy, the Scots Grey is thought to have Game fowl bloodlines, and to have originated from the fowls that once strutted round most of Scotland's farmyards. The Scots Grey is cuckoo-feathered – a most attractive patterning of black and grey that makes it a smart addition to any poultry collection.

Meanwhile, the hen party on the beach was wandering further afield, and had been joined by some domestic ducks. In the distance, shelduck guddled in the wet sand, as the tide wandered further out

Scots Dumpy hen

Scots Grey cockerel

Hebridean tup

across the Atlantic. Donald's pampered fowls were a far cry from their poor, pathetic battery-reared relatives. This glorious location was little short of poultry paradise.

Donald's Hebridean flock was peacefully grazing in a field of orchids and buttercups. Some of the older ewes with gnarled and weathered horns, looked like ancient ethnic women with dreadlocks. Fleet of foot, they galloped round the field as we leant on the gate to admire them. An impressive tup in a field fringed with bog cotton stood and stamped at us, his pointed face framed by wonderful rounded horns – an old breed for new times, now happily seeing a revival in popularity in their native islands.

A small, hardy breed, the Hebridean is a member of the group of short-tailed sheep. With a very dark brown fleece that tends to turn grey with age, their wool is very popular with spinners and weavers, and retains a lovely natural colour with no need for dying. Once the sheep have been clipped, Donald sends the wool to Wales. There it is sorted and made up into balls, and some is returned to Uist to be sold.

Before the nineteenth century the Hebridean was found in many of the outer islands, and the west of Scotland, but after this period it was largely replaced by the Blackface. And confusingly, it had over the years many names including, most frequently, the St Kilda sheep. This has given rise to controversial arguments about the sheep's origins, though there are records of multi-horned, dark coloured sheep on St Kilda before the 1880s. Until as recently as 1979, it was still referred to as the St Kilda sheep, but after this time it was thought more appropriate to call it the Hebridean.

Once it had been ousted from its native homeland by the Blackie, a few small flocks were retained in parks in the South of England. They were an attractive addition to the landscape, not only with their dark colouring, but also because some of them had four horns. This unique trait is still favoured by a few breeders. Unfortunately, many of the four-horned animals also have a tendency to a split eyelid – an undesirable characteristic that has led to selective breeding of animals with two horns.

The Hebridean, despite its small size, may easily be crossed with heavy breeds, and produces a good commercial lamb. The meat from pure-bred animals is gamey in flavour, and almost cholesterol free – a desirable quality in times when we are all so conscious of food scares, additives, and keeping fit. The ewes usually give birth to twins, or a single lamb. Triplets are a rarity, and something most farmers prefer not to have in any case, for there are few sheep that successfully make a good job of rearing a third lamb unaided. Great mothers, they often hide away alone to give birth. In recent times, the Hebridean has been used in special areas of conservation importance, as they are adept at eating scrub and purple moor grass, both undesirable on heathland, and heather moors.

The most primitive of all British sheep breeds, the Soay, has been described as a *living fossil.* As wild as a hawk, it will not be herded traditionally with dogs, and can run like the wind that sweeps over the isles of its forbears. Donald Ferguson's small flock originated from two lambs brought back from St Kilda in the rucksack of a doctor many years ago. This tup and ewe lamb formed the basis of his present flock which is, like the sheep on the island, very interbred. They were losing their wool when we saw them. The Soay is not clipped like other sheep as it sheds its fleece annually, and was traditionally *rooed*, or plucked by the islanders, and the dark brown, fawn or black wool used for weaving and spinning. Many of the animals will have the pale underbelly markings, and patterning of the mouflon. The ewes can be horned or polled, while the tups are mostly horned.

St. Kilda consists of four main islands: Dun, Hirta, Boreray and Soay, and imposing sea stacks – Stac an Armin, and Stac Lee. All are made up of sea-girt cliffs: great bastions of rock that take continual hammering from their exposure to the Atlantic Ocean in all its tempestuous moods. Keith had spent a considerable amount of time on St Kilda some years ago, and had returned with impressive artwork of the landscape, the birds, and the sheep. Sadly the nearest that I came to the St Kilda group of islands, homeland of this capricious animal, was the Island of Berneray, where I had to content myself with a fleeting view of St Kilda one hot July day from Berneray's West Beach. The sea was

Soay sheep on Hirta, with the island of Soay behind

turquoise blue, and squadrons of brilliantly shining, white gannets patrolled the water on their fishing trips from their breeding grounds on St Kilda. Here the world's largest Atlantic gannet colony amasses every breeding season on the stacks and cliffs of Boreray. St Kilda is also home to Britain's largest and oldest fulmar colony, and has the largest colony of British puffins. With more than a million birds, this is the most critical seabird breeding ground in north-west Europe. The St Kildan people were very dependent on the birds for their survival. Alas, they were all evacuated in 1930. Had I been fortunate enough to reach this remote, and fascinating destination then sadly there would have been no St Kildans for me to talk to.

On a clear day from Berneray, the rocky isles that make up St. Kilda appear as strange shapes shimmering on the horizon. Approximately 50 miles due west of Harris, reaching them can be hazardous. Even in a fairly large boat, the sailor is always in the lap of the gods, at the mercy of the swiftly changing weather.

Soay sheep are descended from primitive domestic sheep that came to Britain from Europe approximately 7000 years ago. Today's Soays have evolved from a feral flock that lived on the St Kildan island of Soay, and probably had Viking origins. Over the years, animals were caught up and brought back to the mainland where small flocks were subsequently established. After the evacuation of the island's people, some were transferred to the largest island in the group, Hirta, and have been virtually untouched ever since. Even today, the Soay sheep found on the mainland have retained many of their wild characteristics. Still bearing a close resemblance to its ancestor – the mouflon – the Soay will test the patience of the best shepherd, and take not the slightest notice of his dog.

The St Kildans filed their dog's teeth so that no harm would be done to the sheep as they grabbed hold of them. Sometimes the dogs would also have their canine teeth removed to avoid inflicting damage. With a divide and confuse mentality, the sheep scattered in every direction when attempts were made to drive them. Many would have been lost over the vertiginous cliffs. Others would have escaped capture as they scrambled away from their pursuers with goat-like agility. Clearly there was an element of sport involved in St Kildan shepherding techniques.

I wanted to speak to someone who had been on St Kilda for a considerable period of time, who could tell me a little more about this fascinating, ancient breed. Wally Wright was warden for the Nature Conservancy Council on St Kilda for seven summers. He remembers the Soay sheep fondly, and in particular mentioned the attractiveness of their lambs. During the time that he was there they tried to tag at least 50 lambs every summer. This task was carried out within two days of birth, as lambs had to be caught unawares. Any later they were far too swift and flighty to be caught. The tagging enabled them to find out how long the sheep lived, and also gave vital information about their movements across the island. An experiment was carried out, and some of the males were castrated. The following year it was discovered that a far larger proportion of the castrated males had survived the winter than the tups. This proved that the rigours and stress of mating take a heavy toll on the animals.

During really harsh and stormy winters, the sheep numbers tended to crash. Interestingly, during the summer months the islands are very green and grow good grass. This is due to the huge numbers of seabirds whose guano acts as a rich fertiliser. The sheep are able to build themselves up over the green months of June and July, and can then withstand the severe winter weather.

Hungry seamen desperate for some meat once struggled ashore to shoot some of St Kilda's sheep. Dicing with the huge seas, and struggling over the dangerously acute rock formations that make up this wildest of Scottish habitats, it is hard to imagine how they managed to return to their boats with their stolen booty.

While Wally was on St Kilda, much of the Soay wool was collected by the National Trust work parties that came out to the islands. It could be found rubbed off on the old ruins and dry-stane dykes.

Boreray sheep and gannets

Catching the sheep had long since been abandoned as well-nigh impossible. The traditional old stone *cleits*, once used as food storage places by the islanders, were often a place where the sheep would shelter from the bad weather. Sometimes when the wind was in the right direction, and stealth was used, people were able to corner them unawares. However, a startled beast was not averse to taking a massive leap at the captor as he came to the doorway, springing out with the velocity and agility of a stone from a catapult. There were many tales of bumps, bruises and bloody noses, for the power of these lithe and agile animals, especially when horned, was surprising.

On Boreray another Scottish sheep breed still survives as a feral flock. Now officially recognised by the Rare Breeds Survival Trust, the Boreray, a primitive breed, is a descendent of the Dunface, or Old Scottish Shortwool. When the St Kildan people left the island a remnant population of their sheep was left on Boreray. They had been crossed with Blackfaces and have since had no influences of any other blood. Smaller and more wiry than the modern day Blackface, the Boreray, like the Soay, has adapted to the harshness of the climate, with its numbers fluctuating in good years and bad. A handful of animals were removed from the island in the 1970s, in order to establish some small flocks on the mainland. However, on Boreray, approximately 400 animals still live a totally wild existence today, sharing their lonely landscape in summer with approximately 60,00 garrulous gannets.

Some would suggest that the hardships that are faced by St Kilda's wild sheep are cruel and unnecessary. However, they are perfectly adapted to their savage habitat, and have become a vital part of the ecology. Without their grazing the flora of the islands would alter dramatically. They are a vital part of Scotland's Heritage, to be protected at all costs. These are unique sheep in a stunningly unique location.

On the second last night of our blissful Hebridean sojourn, we took the van down a long sandy track to the dunes at Solas. Heavy rain woke us in the night, and obviously played havoc with the digestive system of the van, as next day it coughed and spluttered before dying altogether. West Coast rain seeps into everything. The poor ageing vehicle had suffered as a result. While I packed up, Keith pedalled off on his bike across the dunes, eventually returning with an assistant and his tractor. After a bumpy tow across the machair, an asthmatic splutter cleared the engine of salt spray and rain. Grinning from ear to ear our willing rescuer waved us off with typical Hebridean charm, despite the fact that he clearly thought we were two mad tourists to have ventured to such an exposed maritime spot in the first place. We hoped that the van would survive the homeward journey.

On the final night, we stayed out on the open moors dotted with a jigsaw pattern of tiny lochans – perfect habitat for golden plover, snipe, curlew, dunlin and diver. My favourite noise echoed round us as snipe drummed, winnowing their tail feathers as they descended in amorous display. Soft Highland rain pattered on the van roof, and as we were just drifting off to sleep, a sad, mournful call filled the night. Red-throated divers. Their cries epitomised all that is so wild and beautiful about the Hebrides, and perfectly summed up my feelings, and my reluctance to leave the Uists next day.

Red-throated diver

CHAPTER TEN

Clydesdale

THERE is no doubt that the Clydesdale has a special magnetism that draws people like wasps round a jam pot. It has been an integral part of agriculture for many years. Not only is it one of the noblest of farm beasts, but it is also much loved and worshipped by all those who can remember the days when they worked with it, and interestingly, many who are far too young to even recall that era. And due to the steadfast loyalty of these Clydesdale stalwarts the breed will continue despite the difficulties encountered in today's fast world. However, it is sad to reflect that though this is surely one of the greatest work-horses of all time, it has now become yet another rare breed.

The name, Clydesdale, originates from the area of the river Clyde where this horse gradually evolved into the animal as we know it today, but its past history dates back to the 1600s when the Duke of Hamilton brought six black Flemish coaching horses to Scotland from France. The influence of this new European blood was soon put to good use as the local farming fraternity shrewdly saw that the horses they were breeding had a far better potential for work when crossed with the new arrivals.

In the early part of the 1900s the Clydesdale was in huge demand and was exported all over the world. During the years of the First World War heavy horses were also required not only at home to work the land, but also to go to war in France. At that time it was hard for farmers to keep up with the demand for their precious animals. As many as 140,000 Clydesdales worked on Scottish farms, and also kept towns and cities moving, for there they were used as draught animals. Mechanisation has largely been seen as successful progress, though it has totally altered the face of the agricultural landscape, and been the death knell for the traditional farm workhorse. Yet, due to the dedication of a few impassioned Clydesdale supporters, their enthusiasm has been passed on to younger generations, and many valuable bloodlines have been preserved. Magnificent beasts, Clydesdales have always been surrounded by controversial traditions, and many of these too have largely been preserved.

(above) Torrs Triple Crown
(opposite) Littleward Lucinda and her foal Morag, at the Carrick's farm, with distant Ben Ledi

The forbears of the modern Clydesdale were predominately black and bay. Now roan and grey are acceptable breed variations, although the latter is very scarce. White blazes are a normal facial characteristic. The Clydesdale, like many other farm animals, has not escaped the whims of fashion. Today, it is preferable for a horse to have four white socks, making them even more eye-catching when shown. In the past Clydesdales were measured at shows and had to be within the bounds of a set height limit, during the selection of stallions for breeding purposes, height was also taken into account. Now height limits have been lifted and the larger horses are often the most popular.

When Charlie Carrick watches Ben Ledi in the mornings, he can usually forecast the day's weather. Though their farmland is very flat, on a fair day the backdrop of distant hills, including Ben Lomond and Ben Vorlich, is clearly visible. Charlie and his son Matthew farm 225 acres at Thornhill, in the fertile Carse of Stirling, famous for production of high quality Timothy hay. The Carrick's predecessors farmed in the same parish for over 400 years, and like them always had horses in their blood.

During its heyday, the Clydesdale was hugely popular in Central Scotland as elsewhere, but it was here that many famous stallions made their mark, including the infamous Baron of Buchlyvie. This stallion became unwittingly involved in a law-suit following a disastrous partnership dispute between Mr William Dunlop and Mr James Kilpatrick. After a protracted and heated court case it was decided that the horse should be put up for auction. The event took place in the packed market at Ayr in 1911. So newsworthy was the whole affair at the time that there were more than 5,000 people in attendance. The Clydesdale has always been able to draw a crowd. After intense bidding Mr Dunlop bought the stallion for the massive, world record price of £9,500, and Mr Kilpatrick was paid back his half share. It is interesting to think of the equivalent value of this sum today. The figure would be astronomical.

During Charlie Carrick's youth, stallions travelled round to serve the local mares. Then the region was divided up into districts, each with a particular stallion: Stirling, Buchylvie, Scottish/Central, and Doune and Dunblane. 'At that time the stallions were licensed by the Department of Agriculture, and the farmer paid a fee firstly for the service, and then subsequently once the mare was in-foal. For example it might be £3 for the mare to be covered, and then about £5 for the foal, although the Scottish/Central horse was usually more expensive as it was often a champion stallion. I remember the family paying £10 for the service, and then a further £10 for the foal sired by the Scottish/Central horse. Much of it was the fashion of the time. The stallions were only meant to travel round their own particular area, but there were also poaching horses that you could use on the cheap. These poaching horses always caused a great deal of trouble and people became quite heated about their activities. Grenadier was a well-known poaching horse. I think he probably left many good foals. In the pedigree of every champion you will always find some unfashionable blood,' Charlie laughs. Currently, an average Clydesdale service fee costs between £80–£100, and then an additional payment of £80–£100 will be paid once it is confirmed that the mare is in foal.

Today, the Carricks only have three Clydesdale mares and foals, but though they no longer work with their horses, they cannot envisage a day when they would not have them on the farm. Showing has become one of the main purposes of the modern Clydesdale, and fulfils a vital role. The Royal Highland Show at Ingliston, Edinburgh, is the highlight of the showing year. Charlie first showed there in 1953. Then it moved round Scotland to a different venue every year. 'It's really not as good now as it was as we are hampered by so many rules and regulations. We used to just sleep on mattresses on the floor in the boxes, and would have our breakfast in the Herdsman's Bar.' Charlie explains. 'Aye, and even in my time I've noticed that there are not the same characters any more, however, for us it is really a bit of a holiday and we look forward to it,' Matthew adds. 'And we'd both rather be showing than judging.' Despite the changes, the four-day event brings exhibitors from all over the British Isles and spectators from a good deal further afield.

For livestock exhibitors, preparations start many months beforehand as decisions are taken on potential show winners, and pampering, primping, and diet play their part in the preliminaries. Today, some stockmen stay in caravans and floats on site, while others chose the purpose-built, but basic stock-attendant's boxes. Much of the show camaraderie centres round the stalls and pens housing the animals, and of course the venue's numerous watering holes. While sorrows may be drowned after a particular beast has failed to make the grade, celebrations of a championship will inevitably end in the compulsory sore head, when next morning, however large the hangover, the breeders and stockmen will have to muster themselves once more for another punishing schedule. And often a good dram will help to soothe the ire of that all too familiar character found at all shows: the wronged owner, who everyone said should have won, but by some odd quirk of fate, didn't.

Once the decision has been taken on the horses and foals that are going to the Highland Show, even greater steps will be taken to protect the feathers round the animal's huge feet. Thick, straight, silky hair is a vital part of a Clydesdale's assets and should ideally be white from knee to hoof flowing sleekly as the animal is trotted out. In the winter with the depth of mud churned up by feet as large as dinner plates, and in our increasingly wet summers too, it can be very hard to protect this valuable commodity, and all too often hair falls out.

During the summer, the Carricks use leather boots, sometimes called spats, over their mare's feet to help protect them from the weather. However, as Matthew adamantly says, 'There is frequently more damage done with boots than without them as they must be checked regularly to ensure that the hair underneath is not matted up, otherwise it will also just fall out. Some people stick boots on and forget them, and of course that ends in disaster.' Though hair is a very key part of a Clydesdale's finery, Charlie laughs about it. 'You will hear people referring to some animals as being "soft of its hair", and then you will also hear them saying, "hard of its hair," what on earth do they mean? It's all over emphasised, it's the quality of the animal's bones that makes the hair.' Blistering, now illegal, was an old practice that was supposed to encourage copious hair growth. Matthew delves into a shed and returns with a bottle, with a yellowing label that reads as follows:

Hilston's Blistering Oil

An excellent application for young horses for promoting growth of hair and spreading of hoof.

Rub well in with fingers round the hoof heads; after the 2nd or 3rd day keep the parts carefully greased with lard to prevent the skin from cracking.

Repeat application after 4–6 weeks.

Jim Aitken

This potion, supposedly based on extract of a particular Spanish biting fly, caused a violent reaction and literally blistered the skin. Sulphur and pig oil was often applied to the affected area to stop it from festering. The end result was a profusion of hair growth, although sometimes things could go badly wrong. Blistering was widely carried-out in most Clydesdale circles before being banned in the 1970s.

'Clydesdales must have a naturally good foot too. Our foals are first shod at about six weeks of age. We always put shoes on any horses we are showing. It is alright for them to be barefoot for the local shows, but for the big shows they must be shod,' Charlie tells me. Clydesdale shoes are different to those used for other horses and ponies. Specially designed, they help to give the horse a closer action. The farrier pares a little off the inside of the foot to encourage outside growth. This in turn slightly adjusts the way the horse moves. Action is another important factor in a good Clydesdale, and the animal should flow fluidly as it trots. From behind the hocks should be very close together, almost touching. 'It's not attractive to see them going wide behind, and we find that if they are close when they are born, they will remain like that, although through age they do tend to widen a little. I remember once at a show one of the fillies in the line-up was very wide behind, and so she was put near the bottom of the class. The owner was not too pleased and later tackled the judge asking why she had been put down. The judge said it was because the mare was too wide. At this the owner looked at the judge who had bandy legs and said, "and so are you". That's showing for you,' laughs Charlie.

At dawn the washing bays round the livestock pens at the Highland Show are a buzz of activity as animals, including cattle and sheep, are shampooed and blow-dried, polished, trimmed and groomed to perfection. The noise of hand and industrial hair-dryers, mainly used for the cattle, fills the air with a drone like a squadron of irate bees. Much of Clydesdale showing is about presentation. Though the horses will have been thoroughly washed at home prior to coming to the show, more shampooing seems vital. Delfoam soap is used not only for washing, but also for pushing the coat up into small waves which when cleverly done can help to mask minor defects in a horse's conformation. The soap is applied dry but care must be taken with the weather. 'If it is wet, then you can end up with a real mess. *Soaping*, as it is termed, is not allowed at the Highland Show. However, it is done for many other shows and can make a real difference to a horse's appearance.' Matthew explains. Once the horses are dry, the legs will sometimes be bandaged to help make the hair lie in the right way. A special powdered resin was once used to put on the fetlocks, but it is no longer made and baby powder has taken its place. 'There is a real art to Clydesdale show presentation, and some people are exceedingly gifted at bringing out the best in their horses. It really can make all the difference,' Matthew says. My mind flits idly to models without make-up.

The horses stand long-sufferingly while their intent owners carry out the meticulous ablutions. Despite the size of the hangover, this is an integral part of show performance. All the different animals at the Highland Show are found in sections according to their breed. Often you may even find that you have been allocated a pen next to your arch-showing rival. The intense titivation that accompanies the Clydesdale has to be seen to be believed. It even continues as the beast makes its debut in the show ring, as a specialist coiffeur follows in the wake of each competitor frenetically adjusting and perfecting during every available moment, haloed by a haze of baby powder. Baskets of sawdust accompany many horses into the ring and are used to dry and enliven the feathers. Traditionally a *spale* basket made of hazel was used to carry the sawdust. Some competitors still own these coveted receptacles but sadly the place in Ireland where the original ones were made was burnt down and they have become as scarce as hen's teeth. However, a revival in the making of oak swill baskets fills the gap. 'We now use a wooden box that does the turn I suppose, but it's not the same as the spale baskets.' Charlie says wistfully.

Winning any class at the Royal Highland Show has much kudos attached to it, but the ultimate goal for the Clydesdale is the Cawdor Cup. This trophy came into being at least 100 years ago. Every year there are two Cawdor Cups presented, one at the Glasgow Agricultural Society's Spring Stallion Show for the male champion, and one at the Highland Show, for the female champion. For the cup to be won outright it must be won by the same person on four different occasions, with four different animals. In 1950 Colonel Medcalf presented cups for the breeders of the two Cawdor Cup winners. This is also a great accolade, but unlike the Cawdor Cup has yet to be won outright. The Carrick's horses have won the Cawdor Cup twice.

In 1987, my stepfather, Michael Thomson, started the Phesdo Clydesdale Stud at his home near Fettercairn, Kincardineshire. Run by Jim and Marjorie Aitken, it was quickly established and put on the map through their numerous breeding and showing successes. In 1993, their mare, Parcelstown Belinda, won the Cawdor Cup. Traditionally, the winning owner is bound to provide celebratory drinks for all the other Clydesdale owners. My mother and I located my stepfather in the Clydesdale Society tent amid the thronging crowds. Following a little too much liquid refreshment he was in dire need of some solid fare. We battled our way circuitously through the packed food hall in search of a plate of sustaining mince and tatties. However, we succumbed to a large plate of oysters instead. While they did indeed provide fortitude I do not think their aphrodisiac properties were put to the test. A Clydesdale man's affair with the Cawdor Cup is a deadly serious one, and perhaps a little exclusive.

Another of the high spots of the Highland Show is the Grand Parade that takes place on the days after the judging is complete. Usually all the winners are surprisingly well-behaved, but even if there has been a great deal of practice put in at home prior to the event things can go wrong. Once one beast misbehaves then there is often a domino effect as other animals start to play-up and the long spectacular lines of prize-winning stock lose their concentration. While it can be hard to hang on to a cavorting calf or foal, the problem is greatly magnified when several hundredweights of prime beef take off on the end of your rope. Pipe bands and the general razzamataz of the show can be enough to cause a skittish attack. For those with a grandstand view, these little flippant moments are all part of the fun, and miraculously rarely cause any real trouble.

Like Charlie Carrick, Jim Aitken grew up surrounded by Clydesdales. On a small farm in the Howe of the Mearns, the landscape for the well-known books of author Lewis Grassic Gibbon, his family relied totally on their horses. He gave up horse ploughing many years ago, but recently was reluctantly persuaded to compete again despite feeling too old for the job. On arrival at the ploughing match he met the two geldings he was going to be using for the first time. Despite having never worked with him before, Danny and Sammy ploughed most beautifully and Jim swept the board, winning not only the prize for the best ploughing, but also the prize for the best turnout. Clydesdale decoration is another important feature. Mares and stallions will often be decorated by the addition of coloured raffia, or segs in their manes and tails. Diamond rolling is the specialised plaiting of coloured cloth, ribbon or wool that is also added to the hair. Some of the traditions of horse decoration have been passed on from father to son, and are in themselves an art form, meticulously hand-crafted.

Anyone who has ever met Jim will appreciate that he is fiercely competitive. He will probably never be too old to compete. With a wry smile he admits that he has always liked a good competition. 'I've aye shown, and I've aye plooed,' he laughs. He has worked with horses for as long as he can remember and as a child would return home from school to follow the plough. He won his first ploughing competition in 1947 and soon became totally addicted to it. In those days there would often be as many as 30 pairs of horses in a competition.

Both Jim and his wife Marjorie left school at the age of 13. After this Jim went straight into full-time horse work. There were usually four working horses on the family farm and they were used for every task carried out by tractors today. The mares worked right up until the time they foaled, and started again with light work sometimes as soon as a week after. Horses were stabled during the winter, or kept in stalls and turned out to grass towards the end of May. They would usually be brought in again once the binder started to work in August. Feeding them was one of the first tasks of Jim's day and he rose at 5.00am, giving the horses plenty of time to digest their food before their work. After the feeding routine, farm children had numerous other chores to complete before they walked to school every morning. There were hens, pigs, and calves to feed, as well as cows to milk. At about 6.00am, the horses were meticulously groomed, the harness was cleaned if it had not been done the previous night, and then the animals were yoked ready for work soon after 7.00am. Rural life revolved round the horses, and, amazingly, the entire rent for a small farm could often be met with the proceeds from the sale of the Clydesdale foals.

Jim and most other farm hands rode their horses out to the fields and home again at night. They also rode them to the farrier. Though many hoof repairs were carried out at home, all the horses wore special shoes with heels that gave them extra grip for their work, and these were regularly replaced and checked. The vet seldom visited the farms. Linseed oil was used for many minor ailments, but the recipes for the cures for numerous others will long remain a mystery, as will the mixtures used to keep the harness immaculate. Most small villages had a local saddler where harness was made and could quickly be mended. Care of the harness was an important task and was not to be neglected at any price, even at the end of a hard day's toil. The horses themselves were often taken to a nearby burn or river where they could paddle in the water to cool off and have the worst of the mud cleaned from them.

Work horses always had their tails docked, usually when they were about three years old. Jim Aitken, Charlie Carrick, and many other traditionalists like them feel that the banning of docking was the worst thing that ever happened to the Clydesdale. They were docked to keep the tail from being damaged if it accidentally caught in the working harness, equipment or reins. Heavy horses are still docked in Canada, although the practice was made illegal here during the 1940s.

By about 1934 the dreaded equine disease, Grass Sickness, was rife all over Scotland, and hundreds of horses died with no cure in sight. Sadly, despite the advance of veterinary science, this still remains the case today. As tractors such as the wee Grey Fergie came in, the horses were soon replaced by them, and were eventually phased out altogether. A few dedicated farmers hung on to them as long as they could. They sadly soon realised though that from a working point of view, the horse simply could not compete with the tractor and new machinery. It was a losing battle. For men like Jim and Charlie this was a sad day, for nothing could replace the intense relationship built up between man and horse. Without doubt it is their loyalty to this unique breed that has kept it and its traditions alive today.

Though we will seldom see the Clydesdale at work in the fields again, the huge power and accompanying gentleness of this tremendous horse will ensure its eternal place not only in the very hearts of Scotland, but also world-wide.

CHAPTER ELEVEN

Luing

ISLANDS have always had a special magnetism for me, and Luing is no exception. Lying south of Oban in a small archipelago in the Firth of Lorne, it is approached from Seil, via the Bridge over the Atlantic. We set forth on May 1st in glorious sunshine, spring unfurling before our eyes. Luing is reached from Cuan Ferry and a small car ferry which plies its way across the sound. Our ferry was due to leave at 1.30 pm. We arrived in good time and sat optimistically on the jetty, watching the little boat crossing the short stretch of water. No sooner had she arrived on our side than she was off again, sending spray from her bow as she retreated like a dog with its tail between its legs. Empty. From the other side of the water, voices carried; we could hear hammering and laughter, then all fell silent. We watched through our binoculars a group of people who sat clearly enjoying the sun. A shag flew past. Meanwhile on our side, several delivery vans rushed down to the jetty; overweight drivers with descending jeans lugged out boxes, puffing and blowing as they staggered down to unceremoniously dump them. One man bent too strenuously, briefly revealing a crack almost as large as the Grand Canyon, hoisted up his breeks, then revved off leaving a fridge stranded in its cardboard box. Still the ferry did not return. Two ladies emerged from their dog-laden car, tutted loudly, then got in again, slamming the doors with irritation. And still we waited.

By now a large black cloud was hovering somewhere above Luing and I was feeling frustrated in case we had several days of rain and did not manage to achieve our aims. Such are the vagaries of West Highland weather that one becomes conditioned to rush about frenetically whenever there is a blink of sun to ensure that one wastes no opportunities. I felt like an irate bull pawing the ground impatiently, desperate to reach our destination and set cameras and sketchbooks in motion. The ladies got out again, tutted some more, then started up their engine and stalled as they tried to do a u-turn on the jetty, finally driving off in a cloud of furious exhaust fumes. It was at least an hour before the ferry returned. Clearly there were technical hitches. Living on an island does have its disadvantages.

We did a whistle stop tour on arrival to absorb the overall picture of the island. It was breathtaking. By now the black cloud was emptying itself somewhere out over Mull, and sun was again painting Luing. The verges were resplendent, thick with primroses and celandines, the trees just showing the first of their summer dress, young oak leaves softly golden, blending with the yellow of the flowers. In one wild roadside garden bluebells and early purple orchids dotted the grass. Feathery grasses grew out of a decaying fence post, and in a bed of phragmites a sedge warbler sang.

Cullipool is Luing's nerve centre boasting one shop and the Post Office, the latter housed in a tiny wooden shed. Many of the cottages were built for quarrymen in the days when the island had a thriving slate industry, and though the last slates were shipped out in 1960 there is still much evidence of the quarry's importance. Now the old abandoned slate workings brood quietly over this attractive little village, whose maze of walls are made of beautiful, near-black stone. The harbour wall too is fashioned from the same material, its stones slotted together in an exotic, abstract snakeskin pattern.

However, it is the famous cattle that have put this tiny island firmly on the map; their story is a truly remarkable one. Specially developed to thrive in exceedingly wet parts of Scotland, the Luing is the first new breed of cattle of recent times, and has, without doubt, been an outright success, proving highly popular not only on Scotland's wet north western fringes, but also in many other corners of the world.

The Luing was the brainchild of the highly respected and astute Cadzow brothers: Shane, Denis and Ralph. Three great agriculturists, and unsurpassed stockmen, their careful breeding of this supreme beef cow has been recognised world-wide. Each of the three brothers farmed in his own right on productive arable ground on the eastern side of Scotland. They all had the same aims, and each had found that it was expensive and unreliable to have to buy cattle in from other farms for fattening.

In 1947 they jointly purchased part of Luing, and shortly afterwards acquired much of the rest of the island. During the 1800s, Luing had been owned by Lord Breadalbane. The Cadzow's aim was to breed cattle on the west, and then send them to their farms on the east where they would be fattened for the butcher. They wanted to develop a particular type of animal that was acclimatised to the incessant wet, and would have inherent resistance to the health problems associated with areas of high rainfall, and above all be able to live outside in winter. Breeding their own replacement cows was also a very important consideration.

They had been buying the highest quality Highland cross Shorthorn heifers, and it was some of these cows that formed the basis for their new herd. The Beef Shorthorn, frequently nicknamed *the great improver*, has great fleshing capabilities. It has been used all over the world and the results of an infusion of its blood in various breeds has done much to improve beef productivity. When crossed with the Highland, with its tremendous hardiness and ability to survive on the roughest of herbage, a highly efficient hill cow is produced. Eventually, the cows were crossed back to one particular Shorthorn bull, Cruggleton Alastair, bred by the renowned cattle breeder, Bertie Marshall, from Wigtownshire. This bull was not only of the old-fashioned 1940s type, but also had a character that would leave a very distinctive mark on his offspring. The breeding proved a great success, and two sons from this liason, Luing Mist and Luing Oxo, were kept and used on their half-sisters. Through the combination of the Cadzow's careful monitoring, excellent stockmanship and knowledge of in-breeding and line-breeding, a type was eventually fixed with all the qualities that they were looking for. The Luing was to be a commercial breed, bred specifically to be self-replacing and a producer of prime beef.

Today the farming enterprise is continued by Shane Cadzow's youngest son, Shane, and business partner Bruce Young. Shane and his wife Tootie have adapted well to their island existence. Though frantically busy with all that is entailed with the paper mountain (sadly now a compulsory part of

farming in the new Millennium) they devoted large amounts of their precious time to giving us a wonderful stay on Luing. On arrival at Ardlarach on the southern end of the 3800 acre island, threatening clouds were gathering in a sky of darkening blues. We stopped to watch a female hen harrier drifting over the rushes at the end of the Cadzow's drive. Then Tootie took us off across the hill roads straight away, to show us the animals and the views at once, for fear the weather might close in next day.

Luing is not a rugged island, and much of the land is of an upland nature. Hugely improved over the decades by the introduction of large quantities of lime and phosphates, the moorland herbage has been vastly sweetened, thus encouraging growth of better grasses. The Luing winter had been wetter than ever and the vehicle slid about as we crossed the hill ground. Large groups of cows and calves stood and watched us back-lit by the evening light, while behind them a pattern of islands unfolded: the Garvellachs, Mull, Lunga, and the tiny Holy Isle – spectacular.

The Luing is a predominately deep red coloured cow with a mossy coat though there may be some colour variation within the breed, throwback to its Highland and Shorthorn ancestry. Roans, red and white, and paler colours are all acceptable breed variations, but the deep red is now largely the favoured colour. Luings are not naturally polled, although a few lines that came through with an imported polled shorthorn bull are hornless. The calves are usually disbudded while they are young. On the island, the cows calve in the spring, usually without any difficulties whatsoever. Luings have an exceptionally high fertility rate, and great longevity, and despite clocking up high mileages, have a reputation for being reproductive well into their dotage. The Cadzows have approximately 550 breeding cows and their followers, and at least 40 bulls of varying ages. The uniformity and great condition of the beasts in their various large groups was particularly remarkable and fascinated me. Even after one of the wettest winters on record, these beasts exuded good health.

The island of Scarba is also rented by the Cadzow family, and, visible from the windows of their house, looms less than 3 miles away between the infamous Gulf of Corryvreckan and Jura. A large number of sheep and cattle are sent to Scarba to graze, and during the winter are fed with special automatic feeders, topped up during regular visits by the stockmen. One of their heifers managed to swim over to the mainland from Scarba, and deer have frequently been known to cross that stretch of water.

Donnie MacKenzie was farm manager on Luing for 30 years and remembers the day in 1965 when the cattle were finally recognised as a breed in their own right by the British Government, following the passing of a special Act of Parliament. The accompanying press report read as follows:

> Following a prolonged study of the breeding performance of the cattle developed over the last 17 years by Messrs. Cadzow Bros. on the Island of Luing, Argyll, the Secretary of State has accepted the recommendation of the Licensing of Bulls Advisory Committee that the Cattle should be recognised as a distinct beef breed. The Breed will also be recognised in England and Wales.

I found Donnie painting a small boat near the shore close to his house at Toberonochy, the island's other small village. Attractive cottage windows had cheery geraniums on their sills, a collie or two lazed about the half open doors, and the bulbous buds of wild fuchsia on the verge of a colourful explosion swayed in the sea breeze. 'We had to pull her out of the water as she was beginning to get a whisker,' Donnie explained. The offending *whisker* of green weed had been rubbed off and the wee boat was now being painted a smart sky blue. Once we had discussed the marine titivations, we retired to the house. 'My job was never boring and the Cadzows were great men to work for.' At this, Donnie proudly took out a letter written by one of the brothers after a show and sale at Oban. In it, he thanked Donnie for all his efforts with the cattle, and said how proud he was that they had looked so well. Donnie's good stockmanship had not gone unnoticed for he was not only awarded a long-service medal, but also received the MBE for his services to agriculture.

(overleaf) Cows and calves, Luing, with Mull behind

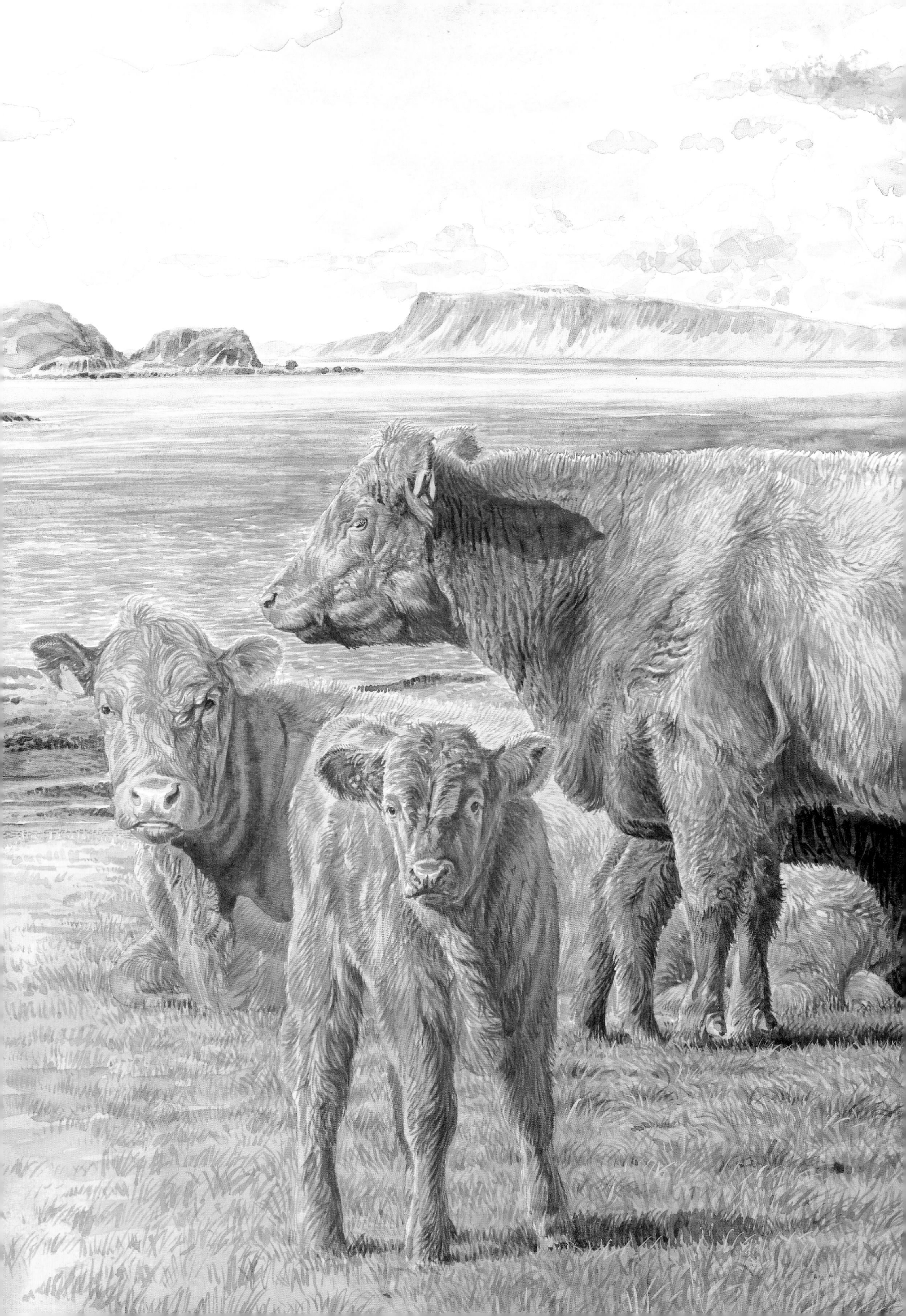

A pile of wonderful black and white photographs unravelled the Luing story. In one a large group of happy souls were wetting the Luing's heads at an island open day to celebrate the birth of the new breed. 'The day they were officially recognised was wonderfully exciting, a bit of history really. The first sales were fantastic too and there was so much publicity and press interest.'

Donnie was born on Luing and went to school to learn English. He briefly left the island for a spell in the Merchant Navy, and also worked on one of the Hebridean steamers, ferrying passengers and animals to and from many of the remotest islands. 'We would start in Glasgow and from there travel to Colonsay, Mull, Coll, Tiree, Barra, Uist and Skye. Cattle were sometimes slung on to the boat and put in pens on the deck. I remember one particular incident when we were on Islay, and were just about to pull out to sea. Suddenly an old lorry came down the jetty and reversed up to the boat. I expected a quiet beast or two to walk on as normally happened, but instead several wild Galloway bullocks shot out of the lorry and crashed on to the deck. They flew straight past everyone and fled on to the upper deck. There were three firemen on board, and they fled too and got themselves jammed behind a wee door and were struggling to hold it shut while the mad beasts were going berserk. It was quite funny really as suddenly a wee head appeared out of a porthole and started shouting for help. Meanwhile one of the beasts had gone into a cabin, knocking a heater over and bending the metal bunk, and then there was all this skittering and roaring. I was very relieved when we finally delivered them to their destination, and what a mess there was to clear up.' I could imagine the purging effect the whole escapade had had on the bullocks, and the pools of swampy green evidence left all over the boat.

Due to his seamanship skills, Donnie sometimes operates the island's ferry. The ferry we had travelled on, the *Old Grey Dog*, was acquired by the Cadzows to make livestock transportation to and from the island more convenient, but is also often used for part-time relief work by the Regional Council. 'The Luings are remarkably docile and we don't usually have too much trouble putting them on and off boats. However, I recall we once had to put some of them on to an old Ballachulish car ferry with a revolving turntable. The animals all went willingly on to the deck, but as soon as they did, their weight unbalanced the boat and it nearly tipped over. We had to adapt it specially to stop it from couping. We have to use boats for everything here as we take some of the cattle over to graze on other nearby islands. The heifers usually go off to the wee island of Torsa. There they are safe from the bulls, and someone goes over every day to check them. They can just be walked off there at low tide when we want to bring them back.'

Coll MacDougall and his mother Annie are also very much part of Luing. 'My father used to work with the bulls, and I was 6 years old when they had the first sale in Oban Market. I have worked as cattleman for many years now and would not want to be anywhere else. There used to be a dairy on the island and the Cadzows had pedigree Ayrshire cattle. The milk was wonderfully rich – full of buttermilk. It was good then, but it wouldn't be any good for me now, as I have a spot of heart trouble.' Coll used to ride a Highland pony round the island checking the cattle, but unfortunately during the summer months it became hard to catch, preferring to spend the day grazing. 'It used to waste so much of my time, so now I just walk or use the bike.'

Ronnie McLauchlan, a well-known Luing breeder who farmed at Ballachulish until his retirement, was at the first sale of Luings, and at the second sale bought a bull. 'There was so much publicity at that first event, and a huge buzz of excitement. The Cadzows were really excellent publicists, and their PR was tremendous. Without doubt this helped to give the Luing a really good start. I had been farming at Ballachulish for a while and really wanted to breed my own replacements. I was attracted by the fact that with the Luings I could do this. At the second sale I bought Luing Star and was very pleased with the results.' Ronnie built his herd up to 45 cows, some of which he crossed with the Simmental. Today the progeny of this cross are known as *Sim-Luings* and are also highly popular suckler cows.

The Luing was one of the first breeds of cattle to employ a fieldsman. This task involves travelling to farms to inspect animals and make sure that they are good enough to be registered. It serves as a link between the Breed Society and the breeders. Obviously, in the early days of the new breed, it was a vital job carrying a great responsibility. In 1972 the Cadzows rang Ronnie to ask if he would consider being their fieldsman. He accepted with alacrity, and thoroughly enjoyed this new position. In the first couple of years he travelled over 60,000 miles to places as wide apart as Shetland and Devon. During the 1970s, Luing popularity was immense and for a while demand outstripped supply. 'I thoroughly enjoyed my work as the Luing is such a remarkable breed and is so easily handled, an important characteristic nowadays with farms having so few assistants. The Luing may be crossed with almost any breed and you will still achieve a good result. Of course she also has the ability to breed pure.' In the beginning, the Cadzows made a condition that the Luing could not be shown. This was because they felt that 50 percent of prize winning represented the cattleman's ability, and only 50 per cent was due to breeding. It was therefore of no direct benefit to the breed. However, after a few years, pressure mounted and showing was finally accepted but soon stopped as the few breeders found that it was expensive to travel annually to the Highland Show, and their bulls were needed for work at that time.

'One of the most outstanding aspects of the Cadzow brothers was their teamwork. Each brother took a turn to be the Honorary President of the Luing Cattle Society, and they always acknowledged the part played in their enterprise by the people who worked with them. I remember Denis once said to me, "Always remember Ronnie, when you are on the farms, to say thank you to the stockmen. Remember these are the people we depend on." It was sound advice and I have always tried to do just that.'

In 1972, the Cadzow brothers were presented with the Massey Ferguson National Award for Services to United Kingdom Agriculture. Their reaction was as follows:

> No rugby match was ever won without teamwork, and what a team we have. The 'forwards' are our men on Luing and what a pack they are – it is their breed, not ours; the 'scrum-half is the Breed Secretary, unstinting in her interest and determination to pass the message back to the 'three-quarters' – the breeders who have, one-and-all, given us tremendous encouragement and backing, and after a hard match, we have, one-and-all, scored a try and won the match with this most coveted award.

For Shane and Tootie Cadzow, and every other farmer in the British Isles, the minefield of new legislation, laws, and health schemes, tagging rules and transportation licensing has made today's farming little short of a bureaucratic nightmare. Many hours that could be spent working with the livestock are spent filling out countless complex forms. Thus their life on Luing now centres round the new enforced paper maze. Then there are the countless other incidents that make up the day-to-day running of a large and remote livestock unit. Shane did not appear for supper, and while we were waiting the telephone interrupted constantly. He had been called to a burst water pipe and was liasing with a digger. Next day he took me for another livestock tour. Just as we were getting into the vehicle, a shriek from Tootie announced that she had forgotten the arrival of some new tenants from England to one of their cottages. We abandoned our plans, and dashed off with the hoover. While scooping up the desiccated blue bottles with the Dyson, I contemplated the diversity that makes up this rural island-idyll. I giggled to myself at the picture of Shane and Tootie negotiating Cuan Sound in the pitch darkness of the wee small hours, buffeted in a tiny boat in horizontal west-coast rain, following a mainland dinner party. Despite this sort of hurdle, their clear love of the social scene has remained unhampered by the remoteness of their lives. Perhaps the Cadzow's new English tenants might find an obstacle like this more of a hazard, certainly a far cry from the convenience of urban England.

Once the spiders had been given a shake-up and the domesticities were complete, Shane reappeared and took me to see some of the stock bulls. Big impressive red boys, they dozed, lazily cudding in magnificent sun as willow warblers in the oakland scrub serenaded them, and a distant cuckoo called. Wisps of red hair blowing on a barbed wire scratching post were being gathered by the starlings – perfect padding for new nestlings. The birds flitted about with their beakfuls of bovine bedding, looking like little old Chinese wise men with long droopy moustaches. We passed Keith who showed us a pair of newly born roe fawns he had stumbled upon, lying perfectly camouflaged and motionless in the bog grasses. Spring really had arrived.

At dinner on the night before we departed, Shane was absent again. A catastrophe. Having travelled hundreds of miles, the travel-weary new tenants were now faced with a major conundrum. The furniture lorry was too big to fit on the ferry. Silage wagons were being hastily washed out, and tractors harnessed to act as removal vehicles. With the typical style of a tinker's flitting, the new inmates arrived in Luing, their precariously perched furniture sticking out of the trailer. Mission accomplished, Shane returned hungrily to Ardlarach for supper. We were just about to eat when again the telephone rang.

The water in the cottage had failed. The day's hassles could not even be soothed away by a calming bath. Shane, placid, steady and obliging, rushed off again. There was clearly an air-lock, legacy of the previous day's burst pipe.

There is hope that perhaps a fixed link may be possible between Cuan and the island. It would certainly make livestock haulage, and fridge and furniture deliveries a good deal simpler. However, despite growing pressures, the importance of maintaining people in remote rural areas still seems to be largely overlooked. But this is a vital small rural community, with a population of 230, supporting a school of between 20–25 children. And it is also the birthplace of the Luing cow, and her loss would surely be yet another miserable coup for our increasingly depressing urbane world.

CHAPTER TWELVE

Prime Beef in Orkney

DESPITE its remote and windswept location, Orkney is prime cattle country. I had heard a great deal about the island's beef production before our visit, and soon noticed that the lush pastures that we passed were full of large, impressive animals, supremely healthy-looking and of the highest quality. It was plainly obvious too, that the Orcadian farmer is a great stockman.

Skaill Farm is situated on mainland Orkney's west-side, overlooking the sweeping sandy Bay of Skaill. Many of the gales that drive in off the Atlantic are aggressively destructive, plastering everything with burning salt. Lick your lips on a day with just a moderate breeze, and you will taste the salinity of the atmosphere. Farming here can be exceedingly difficult as the constant battering of wind and rain erodes many of the fields and rocky headlands. Yet on a good day, this west-facing farm's unique position is glorious, and its location made even more dramatic by neighbouring Skara Brae, the best-preserved neolithic village in northern Europe.

It was during a particularly violent storm in the winter of 1850 that the vagaries of the climate uncovered this astounding 5000-year-old secret. Inhabited before the Egyptian pyramids were constructed, Skara Brae is a semi-subterranean village and has many incredible features that demonstrate how its occupants actually lived. Today its condition is still remarkable, and it is a Mecca for tourists and archaeologists that flock here from all over the world.

The Davidsons have been tenants at Skaill for 60 years. Their 1000-acre farm also has 500 breeding ewes, mostly cross Shetland-Cheviots, and 270 commercial cattle. However, without doubt it is the 30 pedigree Aberdeen-Angus cattle that are Colin Davidson's main interest.

The Aberdeen-Angus was developed from polled black cattle in the north-east of Scotland at the start of the nineteenth century. The Galloway and Aberdeen-Angus, both hornless breeds, undoubtedly

(above) Skaill Delia

have similarities and obviously have origins in the same early black stock. To begin with they were both included in a privately published Herd Book of Polled Cattle. In 1879, the Aberdeen-Angus Cattle Society was formed, and purchased the original copyright of the Aberdeen-Angus details from the author, and reprinted their own Herd Book. The Aberdeen-Angus evolved to be fast-maturing and better suited to an intensive husbandry regime, while still retaining great hardiness. The early breeders were progressive and forward thinking, and dedicated many years to careful line-breeding and sheer hard work. The end result was the fixing of a highly productive and beefy-type of animal that quickly became established at the forefront of the British meat market. Soon after its recognition at home, the Aberdeen-Angus became famed world-wide and was also established abroad with a flourishing export trade particularly to America, Canada, Australia, New Zealand, Argentina and South Africa. Though for many years the breed had enormous success, sadly this was not to last.

Over the decades the breed has undoubtedly been through a great many ups and downs, and like some others, has had to change in order to keep up with modern livestock farming, and stiff competition from continental beef breeds. Today's Aberdeen-Angus tends to be a much larger animal than its early forbears, but has always managed to retain much of its distinctive characteristics, its incredibly eye-catching appearance, and its glossy jet-black colouring. The red Aberdeen-Angus is unusual. During the 1970s many Angus cattle were imported back into Britain, particularly from herds in North America. The advancement of artificial insemination and embryo transplant techniques has been very important, particularly in situations where live imports of animals have been restricted. It has allowed important bloodlines to come back into the country and breeders to retain the position of this beautiful breed at the forefront of a highly competitive market as the world's premium beef producer.

It was in 1974 that Colin Davidson decided to join the Aberdeen-Angus Cattle Society. 'I had no money at that time, and on reflection it was a good job, or else I would have definitely wasted it. The Angus was going through many changes, and not all of them for the better. So I just had to sit back and watch for a while. Many good Angus cattle had been exported to Canada during the 1930s, and the Canadians were still breeding a type of animal that I really liked. In the 50s and 60s a very wee type of Angus was popular, but it was not for me. It was the Canadian animals that had maintained their size that appealed to me the most. Though I bought a few cows at the cheaper end of the market and tried to improve them, this was very difficult as they had been line-bred during the 60s, and their type was very fixed. I was unsuccessful. So I went away to Australia to give myself time to make some money and think about the future. There I was very well paid working on building sites, and then building bridges in the north. It was very hard work as some days it was as hot as 44 degrees. Food was provided and so basically I did not need to spend anything. When I returned to Orkney I had quite a bit of cash in my pocket and decided that if I bought myself a decent car I would probably just end up wrapping it round a strainer post. Perhaps you will have noticed that we have stone strainers here as we have so few trees.' Colin laughs. 'Instead I bought two Canadian cows and that was the start of it.'

It was then that all the planning began and the founder members of his prize-winning herd were put in calf using AI from a Canadian bull. 'It was very exciting waiting for the first arrivals and one of them was a bull calf. Surprisingly, he made the grade on our first selling trip to Perth Bull Sales, making the fifth highest price of the day. That was a big boost. The following year, I took four bulls to Perth because I thought we were on the right track. However, it was a total disaster. It was the first sale in the new market and it rained so hard that the poor beasts almost floated away out of it. It was a bad sale in general with the Angus' making very little money. While I had thought my animals looked good at home, somehow when I got them to the sales they just did not look up to much. I was newly married at the time, and I think poor Pam wondered what she had taken on, and why we had bothered. It can be daunting setting off from here with the trailer on such a long trek. We do need a calm day as it is much better for the animals on the boat. Though we really love Orkney, there are definite disadvantages to being so remote. However, this failure did not put me off and merely made me all the more determined to succeed. I knew that I obviously had a great deal to learn. It was the second cow that I had bought, Nicks Delia, who did well for us as her daughters started to produce really good bull calves. Then suddenly things seemed to come together. Now all the Angus cattle here originate from those first families.

'Breeding cattle is like selling women's clothes, there's no point in coming with last year's lines this year. Staying ahead of the game is the problem.' Colin is a highly ambitious man with an extremely forward thinking brain. He has planned his pedigree Aberdeen-Angus breeding programme like a skilled army manoeuvre. And it has certainly proved successful as at Perth Bull Sales in February 2000, one of his heifers, Skaill Delia, made the female breed record price of 16,000 guineas. 'Our Aberdeen-Angus herd is my lottery ticket, but the trouble was, I was so exhausted and nervous following a huge build-up to the sale, that I was on the sidelines leaning on a gate feeling sick when it all happened. As we were getting everything ready for the long journey from Orkney to Perth, the ferry company had rung to say that the weather was worsening and that we should go on an earlier boat. Then there was a terrible mad rush and we only just made it. They don't let livestock go on the ferries on a rough day. If we had missed that boat it would have been another week before we could have reached the mainland with the heifer as the weather was so terrible. Then of course we would have missed the whole event,' Colin says.

'After all the hassles and preparation, I was just absolutely punctured. I knew the heifer was exceedingly good, as we had taken her to Thainstone Market at Aberdeen the previous December and everyone liked her. Though I had been offered up to £10,000, I really did want to have a Perth Champion, and somehow had the feeling that perhaps she had a chance. So I didn't sell her, and was terribly nervous in case I was wrong. Perhaps I would get less at Perth, but I was willing to take the gamble,' Colin explains.

'Before the sale, I was nervous too because I know what Colin is like,' continues his wife, Pam, 'I was standing combing the heifer and an old stockman came up to me and studied her for a while, then he said, 'Aye, that beast's at 12.o'clock.' I looked at my watch and said, 'No, no, it's ten to two.' I wasn't used to the term, which referred to the fact that she was in peak condition. I felt such an idiot. Looking back it was very funny.'

Perth has always been associated with the Aberdeen-Angus. The first show and sale of Aberdeen-Angus bulls was held there in 1863, while the Perth Bull Sales have been in existence since 1870, and were initially important for sales of both Aberdeen-Angus and Beef Shorthorn. Perth soon became famed as the world centre for these two main beef breeds, and many continental breeds are also sold there during its February and October Bull sales. Auctioneer, David Leggat, remembers being totally overwhelmed when he saw hundreds of bulls at Perth in 1975. 'It has always been an exciting event. And it is here that a bull literally did end up in a china shop. When the market was still at its old premises on Caledonian Road, an Angus bull took off down the streets. It was very funny because in those days members of the public were quite used to seeing the occasional animal in the town. The bull ended up down the High Street and accidentally, in its excitement, broke the window of Watson's china shop. Eventually he was caught and tied up to a near-by lamp-post until someone retrieved him. Of course if that had happened today with all the health and safety regulations, there would have been a real outcry. But no-one minded then,' David laughs.

'Perth Bull Sales reflects the changes in fashions and trends in the beef industry. For me the most exciting thing about them is seeing the true dedication of the cattlemen. They work so hard and are often out at night calving cows, and sorting out problems. Of course if their animals do not do well then it can mean that they may even lose their jobs, so it is a very nerve-wracking business. Without these unsung heroes and our native breeds there would be no Bull Sales. I actually sold Colin Davidson's fantastic record priced heifer, and it was a wonderful moment as I know how hard he has

Bull at Skaill Farm

worked for success. Orkney has a very short summer and a difficult climate, but despite this they also have supreme cattle. Many of the commercial herds are based on Aberdeen-Angus cross Beef Shorthorn, and then put to continental sires. And it really works. Orkney has some excellent herds of pedigree Aberdeen-Angus. I always enjoy my trips out there. I feel very privileged as in my job I see the country's top livestock,' David tells me.

'Orkney Gold', is a label under which the island's prime beef and other specialised local products are marketed. Much of the Davidson's meat is sold through this scheme. With the rich grassland, and excellent stockmanship, it is little wonder that the island has been famed for its beef production for many years. The Davidsons eat their own home-produced animals. They go to be killed at the local abattoir, and are then hung for three weeks. 'Distinctively marbled and with a delicious flavour, we not only know what we are eating, but can also vouch for the fact that Aberdeen-Angus beef really is the world's best.' Colin is rightly adamant.

With his progressive outlook, Colin Davidson was the first person in Orkney to try embryo transplants. Many of these are implanted in the commercial cows on the farm. It is a very tricky business that is dependant on feeding, the weather, and crucial timing. A non-surgical procedure, it involves bringing the donor and the recipient on heat at the same time using hormone injections. The donor is then served either with AI or the chosen bull, and then after carrying the fertilised embryos for seven days, they are flushed out, and preserved in a special straw in a similar way as semen is collected from bulls. Finally, they will be implanted like AI in the chosen recipient cows. It enables the farmer to have

many calves from a particularly good animal in the same year. The average number of eggs produced per flush is seven, and a 60 per cent success rate is usual. When properly carried out it can be a very efficient and worthwhile procedure. One year at Perth, Colin sold six frozen embryos for £1000 each. Foreign embryos may be sent over to farmers when they are perhaps looking for a particular blood-line that is important to keep a good line going. During the 90s, Colin donated some embryos to Kirkley Hall College where they were working on an Aberdeen-Angus development project, as he felt it was an important venture. 'I have never followed the herd and have always chosen to do my own thing with regard to cattle breeding. I have never used the popular bulls of the time. The most success I have had is with line-breeding. The aim of this is to produce a mould and try to bring out all the animal's good points. The closest we will go is mating a half-brother and half-sister, a bull and heifer by the same sire. I spend a great deal of time planning,' Colin laughs. 'I suppose I am obsessed with getting good results.'

Pam is totally resigned to living with a man who lives and breathes Angus cattle. 'The whole attic is full of all the magazines, papers and pedigrees that he studies for hours on end. Every cow's future is well-mapped out for her, as he memorises pedigrees and blood-lines and how they will suit a particular beast,' Pam laughs. 'Colin is a walking Aberdeen-Angus encyclopaedia.' Intrigued to see what would happen if he used some semen from a 60s-type small Angus on one of his best cows, Colin wondered if this cross would produce a good result. 'I admit it was a total disaster as we ended up with this little, dumpy animal that looked tiny compared with all the others in the field. This reaffirmed my

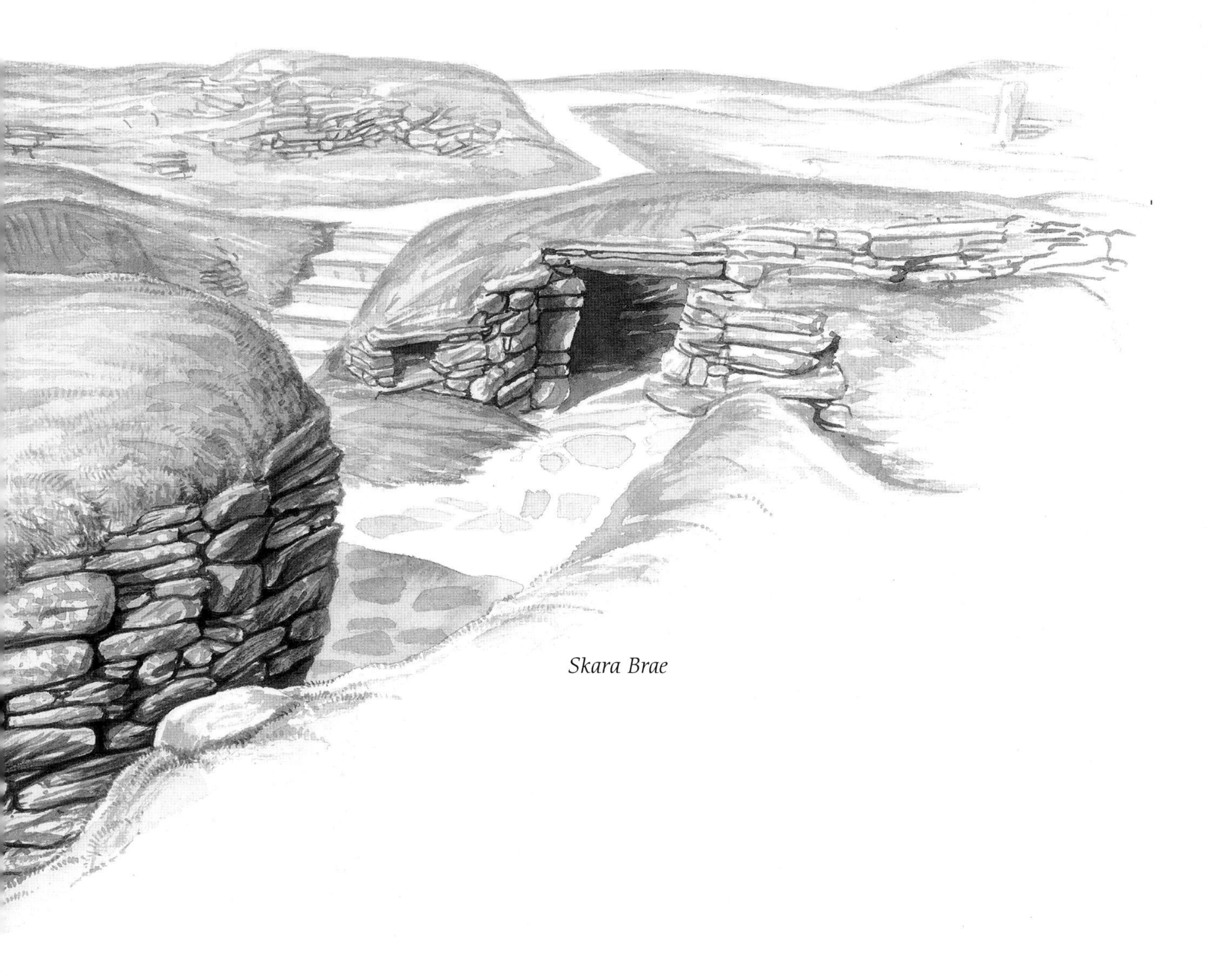

Skara Brae

belief that the old type of Angus had nothing to offer me. Conformation is now very important as the Angus has to compete with the large popular continental bulls, and it is really due to this that they had to get larger in order to hold their own. Without doubt the small animals had their place at the time, but I do believe that we had to move on, and there is still no doubt that the Angus produces the world's very best beef, there is simply nothing to equal it. More Angus beef makes the spec than that of any other breed,' Colin explains.

'I know I have a reputation for being very outspoken, but I know exactly what I am aiming for. Diplomacy is not my strong point, particularly if I have had a few drinks.' Pam listens to her husband and smiles. 'If I am judging, I do not hesitate to put an animal down if I do not think it is worthy of merit. It does not matter what has taken place at other shows, I always like to make my own decisions. I hope that people then know I am always honest.' At this we take off round the fields to see the cattle. Large, shining, deep bodied, and imposing, Colin explains that he likes the cows to really look feminine and the bulls to be extremely masculine. 'They must have plenty of character, not a wee boxy head, and I do like to see large, open ears. If a cow has an ugly head, as far as I'm concerned she will never breed right. Obviously, we don't want the animals to be too big, for the bigger they are, the more obvious the faults.' He laughs. Just coming into their sleek summer coats, the perfect beef-producing conformation is very apparent, showing the Angus' unsurpassed fleshing capabilities. 'They are easily calved, and usually have the most excellent temperaments. As far as I can see, the Angus has a really good future, particularly if we move with the times and keep up standards of good stockmanship, and continue to work with really good blood-lines.'

A silvery trail of milky slaver hangs in the wind like a liquid cobweb as a large black bull calf suckles under the deep flanks of his attentive mother. Perhaps a future champion? Other calves lie in a group together, eyes shut as they doze in the warmth of the sun. The fields are all bound by Orkney dykes, the ditches along the roadsides pink with campion. Bulls of varying ages stand in their field overlooking the activities of Skara Brae as a line of tourists temporarily step back in time. Their future may be a long ferry crossing, an exciting visit to the World's Best Bull Sales, and possibly a farm in Devon. Colin's animals have ended up in many places, and with his growing reputation as a skilled and dedicated Aberdeen-Angus breeder will obviously remain in big demand.

As our visit to Skaill Farm draws to an end, Colin takes us to the far end of the village. From here we can just see the distinctive shape of the Old Man of Hoy, shimmering in the distant heat haze. Hoy means high island as it is the highest part of the Orkney archipelago. Its tallest hill is 479 metres. 'Apart from Hoy, Orkney is so flat that you can see what your neighbour is doing, and as you see it is also pretty windy. I am sure I might have been eight inches taller if it hadn't been for the wind,' Colin laughs. From another viewpoint above the bay, small lochs shine in the sun, a wintering haven for whooper swans.

We drive up a farm track across an eroded piece of headland to the far west of the farm. From here the drama of the red and pink Orkney cliffs unfolds before us, as we battle against an intensely drying wind. Carpets of thrift and sea-campion cover the pitted rock, and in a huge geo, the Hole of Row, fulmars and guillemots nest precariously above the dark inky sea. High on a cliff-face a raven's nest towers above a pair of busily nesting shags, made from dozens of pieces of rusting fence wire, it illustrates that Orkney's birds are equally as innovative as its people.

(opposite) Cow and calf, Skaill Farm, Marwick Head behind

CHAPTER THIRTEEN

North Ronaldsay

OUR flight to North Ronaldsay cost £5. No fiver ever gave me better value. Weighed down with all our gear, we boarded the 8-seater Islander plane at Kirkwall's smart new airport, and took off from the runway with the minimum of hassle. Within seconds of being airborne the views below were simply breathtaking, revealing a turquoise sea and a pattern of Orcadian islands, criss-crossed with dykes and lush fields of prime Orkney cattle. White sandy beaches glistened in the sun, fringed with emerald green weed. Two small boys returning home from a shopping trip to Kirkwall with their mother, were, unlike me, blasé about the scenery. The heavily subsidised plane provides a vital twice-daily link for this far-flung, north-easternmost Orkney island. Remote and unspoilt, North Ronaldsay certainly left a huge impression on me.

Our base for the week was the island's Bird Observatory run by Alison Duncan, the chief warden and ornithologist, and her husband Dr Kevin Woodbridge. Here we were provided with comfortable accommodation and excellent food. The doctor came to the island in 1977, having previously worked exhausting 120-hour weeks in a Manchester teaching hospital. The decision to take up the post as the island's single-handed health care unit marked the turning point in his career. Prior to this, Kevin had always had holidays in Scotland, and having worked for a few months as doctor on the Isle of Lewis, vaguely knew what he was undertaking. With a passionate love of natural history, and a determination to maintain and protect as much of the structure of the rural community as possible, he has thrown himself into this unique island existence with huge enthusiasm.

Alison came to North Ronaldsay on a bird-ringing holiday in 1987, and decided that this was where she most wanted to stay. The couple have two children, Heather and Gavin, and are campaigning to make use of the daily flights to transport them and the few other children of school age to the High

School at Kirkwall. Under the current arrangements, once they reach the age of 12 they must go to mainland Orkney for their senior schooling and stay during the week. 'Basically, all the infrastructure is already in place and it would even be cheaper for the council than putting them up in the hostel on the mainland. It is after all only a 15-minute journey. We just do not want to have our children away from us, and feel that it is important that they come home every night. A service like this would help to keep people with young children on the island.' Kevin explains. Presently, the junior school on North Ronaldsay only has five pupils.

Kevin Woodbridge is a kind and quietly determined man. He has a great dedication to the island, and is modest about his many achievements. North Ronaldsay is four miles long and two miles wide, and in so small a place volunteers can be in short supply. Kevin and Alison, both fit and able, do not hesitate to put themselves forward. They are both in the fire service, though so far a uniform small enough to fit Alison remains elusive, and both are coast guards. Kevin is chairman of the Community Council. He has done all the plumbing and electrical work for the Bird Observatory which was established in 1987. It has been meticulously designed for energy conservation and is largely powered by wind generation, another of Kevin's specialist fields. A large windmill hums away in the background, its stays providing an ideal scratching post for their flock of approximately 100 North Ronaldsay sheep. Amongst the many hats worn by Kevin and Alison, their day-to-day lives also include looking after the various guests that come to stay at the Observatory, providing their meals, and running a small restaurant and bar. As well as keeping sheep on their croft, they grow a little bere barley – a primitive form of grain harvested with a binder and used for making meal. 'I really am in love with my old binder. But after it has lain in the shed for almost a year, it usually seizes up. Invariably, the chains need greasing and oiling, and even then it is always breaking down with the twine coming out in knots. However, when we have almost finished the field, she always purrs very sweetly and leaves a lovely regular pattern. That's what makes it all worthwhile and I do feel it is vital to keep these things going,' Kevin laughs.

Today, a general practitioner is inundated by paperwork. For Kevin as the island's doctor, nurse and pharmaceutical dispenser, the problem is three-fold. He is painfully aware of his responsibilities. 'Often emergencies all seem to come at once, and you can find yourself running round the island in circles trying to keep the situation under control until patients are flown off to hospital. When an accident happens out here it can be exceedingly traumatic and very nasty, and I am all too aware of the seriousness of my position. We have had some tragic incidents and it is then that you realise the implications of remoteness, not only of location, but of being single-handed.' Kevin explains gravely.

But the island's fantastic wildlife and the general way of life have been more than enough to keep the Woodbridge family here. During the migration periods North Ronaldsay is an important stopping off

Dr. Kevin Woodbridge

point for thousands of migrants, a few of them rare. It was due to this that the Bird Observatory came into being, and a net-making business was set-up on the island to harmlessly trap the visiting migrants for ringing and research purposes. Kevin has been ringing birds for much of his life, and though he appreciates the rarities, finds the nights where there is a steady drizzle, mild fog and a south-easterly wind during the migration season perhaps the most exciting. Then literally thousands of birds can be drawn to the lighthouse at the island's northern point. 'I remember standing there on one occasion during what we would call a "classic lighthouse night", in the days before it was automated, and watching perhaps as many as 60,000 fieldfares and redwings. That is one of the things you remember even more than some of the extreme rarities that occasionally land here.' During our time in North Ronaldsay, an icterine warbler and a long-eared owl were caught in the bird traps. Prior to our arrival, a bluethroat had dropped in.

While the bird life of North Ronaldsay is quite awe-inspiring, and often left me totally mesmerised, it is the island's 3,000 North Ronaldsay sheep that fascinated me beyond any other breed. Small, wiry, lithe, colourful, and extremely athletic, I followed their activities with avid enthusiasm, and spent hours pursuing them along the 14 miles of varied coastline, intrigued by their routine.

North Ronaldsay is totally unique. With approximately 12 miles of Grade A listed dry-stone dyke surrounding the coastline, the sheep are kept outside the fields, on the seashore. Through years of evolution they have adapted to survive on their seaweed diet and follow the tides with precision timing, skipping out to the skerries as they are uncovered at low water in search of their favourite seaweed, dulse, and laminaria. It is the pressures of life on the beach that have created this breed. In an extraordinary management system, the ewes are brought in to the crofts from the shore prior to lambing. But the tups and wethers (castrated males) remain on the shore for the entire year so only see a small amount of grass found occasionally between the beaches and the wall. During the summer the ewes and lambs are marked and put back on the shore where they stay until the following spring. Usually a ewe is left with one lamb to rear. Should she produce twins, then the weakest of the pair, usually the female, will be humanely culled. Obviously, too many females means too many lambs the following year, which in turn puts further pressure on the limited grazing. While this may seem surprising, the harshness of the winter on the seashore rarely makes it possible for two lambs to survive. For the wethers, life is a beach and they remain there for several years, before being sent to the abattoir when they are between 3–5 years old. The North Ronaldsay sheep is famed for its succulent, distinctive flavour, its low cholesterol and lack of fat. After a particularly strenuous day's walking I returned to the Observatory to find roast mutton on the menu. It was one of the most delicious and flavoursome meats I have ever tasted.

North Ronaldsay sheep are hardy. With feet that never succumb to foot rot, agility like Russian gymnasts, no need for obstetric interference, and the ability to stand the very worst of the Scottish climate, they are perfectly adapted to their maritime existence. Not only do they behave differently to other sheep, but they smell totally different too, and their distinctive odour bears no resemblance to that of either goats or deer. A few small flocks are kept on the mainland, but have to be given a special supplement as without this they can die from copper poisoning. Seaweed is low in copper and the sheep have evolved to extract enough from it to survive. However, on a diet based only on grass, their systems absorb too much. On the island, their maritime habitat makes them extremely vulnerable to oil pollution. These are no ordinary sheep.

I began my day with a walk along the beautiful beach adjacent to the Observatory. From here the coast turns rocky and is usually covered with the supine bodies of hundreds of seals, both grey and common. With the wind in the right direction I could crawl to within a few feet of the dozing animals. Provided that they did not catch my scent and I kept perfectly still while they were looking at me, they frequently remained on the rocks. Scrabbling about over the barnacles on my stomach and wriggling

oblivious through the occasional salty pool, I was so absorbed by the animals that time flew far too fast. Grumbling and snarling at one another, the soporific seals lay in front of me, the smell of their fishy breath wafting into my face, accompanied by the brief trumpet blasts of their unrestrained gut noises. It was fascinating being so near to this fly-on-the-wall phocine documentary, pelagic intrigue at its best. It was still early June and the common seals had just started to give birth. I came across a newly born pup with its attentive mother and took dozens of photographs from a safe distance avoiding disturbance and conflict. Seal bites are amongst the worst and, rightly, they will not hesitate to put you in your place if you venture too near. With the protection of a few yards between us, I watched as the pup began to suckle, while snipe drummed out over the sea, and the cries of curlews and ringed plover filled the air.

During my walks round the island, the singing of the seals was haunting and melodic. It became an important background sound that blended with the salty breeze, mingling with the clamour of the nesting fulmars that sat close up against the dykes. Care had to be taken not to surprise them while rounding the corners. Fulmar oil spat hastily at any intruder is not only vile smelling, but also so viscous it cannot be removed from clothing, not even with the promise of many of today's miracle biological washing powders. Armed with this lethal weapon, a fulmar can render any threatening creature useless.

From burrows under the rocks and on the sandy banks, the sad little whistling songs of the nesting black guillemot drifted. Nicknamed tystie, this member of the auk family has smart bright red legs, and a dramatic red gape, seen when it calls. Against a sea of aquamarine, small groups of tysties sat singing on the rocks. Dapper and with perfect plumage, they painted a glorious picture as seals languished, and sheep munched their way through long dangling fronds of brown glistening weed.

As if an alarm clock had gone off, as soon as the tide reached a certain level, the sound of hoofs

Incubating fulmar

could be heard, as a long line of sheep started to trek down to the newly exposed rocks. In highly business-like fashion, the purposeful ovine troop bounced lightly over the dozens of seals and carried on to the places where the weed was at its best. The seals barely moved, only occasionally offering a minor grunt of disapproval. Perhaps an eye briefly opened and then shut again quickly as they drifted off in their slumbers. The shining kelp swayed in the swell round the skerries as the sheep agilely jumped out to graze. Though they are accomplished swimmers, it has been known for younger, less experienced animals to venture too far out in their enthusiasm to reach the best weed. Once engrossed in their eating they leave their return too late. Occasionally a few drown as they try to swim against the fast currents of the turning tides.

The dyke system that keeps the animals on the shore has been in existence for many years. It demands a high maintenance input for it is often damaged by the ferocious gales that sweep in off the sea. Frequently the foundations are eroded by the weather, and large areas may collapse, or be undermined by sand, or castings, blown up against the base. As dyking is a labour-intensive task, the island's depopulation poses a problem for the maintenance of the walls. Traditionally, the dyke was equally divided up between the six townships and maintained by the crofters living in each one. This is a wall that would provide enough work to employ the skills of a professional dyker all year round.

While much of my attention was focused on the seaward-side of the high dyke, the landscape on the other side is equally breathtaking. Vast drifts of yellow flag iris surround small lochans, and teal and mallard hide in the thick vegetation, their peeping ducklings appearing fleetingly as they paddle through the water. Redshank, oyster-catcher and curlew busily protect their new young in the waving vegetation. Orchids and buttercups carpet the open areas, while on a maze of lichen-patterned walls, wrens and rock pipits feed their begging young. High above, snipe compete with one another drumming incessantly, rising and falling, filling the sky with their perfect reverberations. Arctic terns scream and wheel as I venture too close to their colony, dive-bombing me with great accuracy, brilliant light shining through their long white wings against a canvas of eternal blue.

Members of the primitive group of short-tailed sheep, the North Ronaldsay have a wealth of colours that blend perfectly with the seascape. Moorits, blacks, tans, greys, and flecked fleeces are but a few of the colour variations. While some of the tups have huge imposing horns sometimes striped with black, the horns of others are so gnarled and curly that they frequently have to be attended to as they grow too close to the animal's faces. There is a huge variation within the breed, with some ewes horned and others polled. Many of the lambs look as if they are covered with beautifully woven tweed, a heather-mixture of earthy shades.

Sadly, in recent times, a sheep's fleece has been as good as worthless. However, like the unique meat of the North Ronaldsay, its fleece, too, is now in demand and those from the previous summer were being carefully stored in one of the lighthouse buildings. Jane Donnelly has grown up with the North Ronaldsay and remains fiercely loyal to the breed. She knows as well as anyone the habits and characteristics of the animals, and has always worked with the family's flock. She and her husband recently travelled to Prince Edward Island in Canada to see a specialised de-hairing and wool processing plant in action. The aim is to purchase this highly efficient equipment to set up a small-scale woollen mill in the redundant lighthouse buildings, processing the island's clip and some from other small producers too. Many primitive breeds of sheep have stunning wool that does not need to be dyed. However, hard guard hairs within the fleece cause some people to steer away from garments made from their wool as they can be scratchy to wear. Jane showed me hanks of wool that had been through varying degrees of the de-hairing process. All were soft to the touch, but the sample that had the most guard hair removed was almost on a par with cashmere.

The project is ambitious and has involved considerable effort to raise the £100,000 required to buy the equipment. However, not only will this bring valuable jobs to the island and put money back into

Ewes and lambs with North Ronaldsay Lighthouse

the community, but provide endless possibilities for unusual garments. Even the removed hard hairs will not be wasted as they can be made into felt, useful for loft insulation, curtain making, upholstery padding, and many other items. Fleeces from other breeds too can be sent for processing on the island. Currently, the Wool Board pays 2p per kilo of wool, but Jane pays up to £1.20p per kilo for coloured wool. She knows its great potential. *A Yarn from North Ronaldsay* will open many doors and help to ensure the future of both islanders, and their sheep.

After visiting Jane and her husband, Peter, I went to see Billy Muir. He had just completed painting the lighthouse, soon to be opened to the public. He and Jane, and other board members have been working together to help raise funds for the new wool venture, and also the Lighthouse Trust. Billy was local assistant lighthouse keeper here from 1969, and today still looks after and maintains the lighthouse. It was one of the last to be automated in 1998, having replaced the island's first tower in 1854. The latter was one of the first four lights to be established in Scotland in 1789, and still stands on Dennis Head. Now its only occupants are nesting ravens on a window ledge near its top.

There are 176 steps up to the top of the North Ronaldsay lighthouse. It must have given Billy a crick in the neck as he wound his way up the tower with a long paint roller. While gazing down from the top of the lighthouse at North Ronaldsay stretched out before us, I became very aware of the vital importance of safeguarding the longest continuous stretch of dry stone dyke in the world. From here the view of the punds and plantie cruzs is spectacular. Every croft had a plantie cruz, a circular area of dyke built to protect vegetables or seed from both the weather and the sheep. The punds are stone built enclosures into which they are driven at pundings, or round-ups. These take place several times a year and have to be carefully co-ordinated to coincide with the tides. The island is divided up into specific areas, each with its own pund. As dogs are rarely used, manpower is important as people drive the sheep into the punds to be handled, sorted, chosen for culling or brought in off the shore. But the sheep are often very unobliging, and with fleetness of foot and incredible agility can avoid being rounded up altogether. Everyone has a story of their cleverness. Some merely take to the water and swim out to sea avoiding the whole issue, while others just leap straight into the air and over the line of people. For the helpers there are frequent bruises and soakings as they try to outwit fleeing animals.

Billy had managed to bring a few sheep into one of the punds near the lighthouse. Several of them

had avoided capture during the last punding, but while checking his sheep he had seen his chance to drive them in so that they could be clipped. He caught a tup as it shot past us in the pund and we compared its body condition to that of the wethers. Under his heavy fleece the tup was remarkably lean. During the hard winter months the animals tend to put on more condition as seaweed is washed ashore in great heaps. Years ago it was traditional for the wethers to be culled at Christmas-time. With the pressures of the breeding business, the tups are always the leaner animals and are not used for meat. They frequently injure one another as they battle for supremacy on the rocks. In the autumn the clashing sounds of their large interlocking horns echoes round the shore. We visited Billy's ewes and lambs in a field adjacent to the lighthouse. Used to receiving a few nuts every day, they came running up to the fence while he pointed out an old ewe that is a particular favourite. Despite diminishing dentition, she will clearly stay on indefinitely. Our sheep tour took us round the island as we visited many fields of ewes and lambs and Billy explained variations within this ancient breed. One characteristic remained the same for all of them. Speed.

North Ronaldsay sheep have always been managed by an ancient Sheep Court. Originally based on the island's six townships, each had two representatives – 12 wise men that controlled the decision-making, with the laird as chairman. Now all sheep owners are on the court and jointly make all the

North Ronaldsay praams with the original lighthouse

decisions on sheep husbandry. Recently the crofters have found that their wonderful meat has a niche in the specialist market, ending up on the menu in a few of London's most prestigious restaurants. This is an area they would greatly like to expand.

During my last morning on North Ronaldsay, I toured the island with crofter Bertie Thomson. Having grown up with the sheep and spent most of his life working with them, no one knows better than him how clever they are. 'Sometimes I swear for them, but they are a real challenge. Especially when they are eyeing you up as they sit on a rock, and won't come ashore until you are well out of the way. They are great swimmers although occasionally some of the biggest ones go too far out in their search for better feed, and do come to grief in the fast tides, but they are as agile as cats on the rocks.' He explained the vital importance of the dyke. 'Sometimes a sheep may learn how to climb over the dyke. We refer to these animals as *loupers*, and unfortunately they must be culled or else they quickly teach all the rest to do it too.' A cold wind was coming straight off the sea as we got out of Bertie's noisy car to study all the different ear marks on the sheep, which correspond to each croft. 'There's one of mine over there, it looks pretty fit, time it was away I think.' Bertie's eagle eye scanned round the headland as he pointed out the attributes of different sheep. They are very territorial and tend to remain in certain areas of the island, similar to the way in which sheep are hefted in other parts of Britain. *Clowgang* is the local name given to the area in which a group of sheep remain.

Bertie pointed out some old boats on the rocky shore at Dennisness. These traditional North Ronaldsay *praams* were very versatile. Built on the island, they were specially suited to shallow waters and going over rocks close to the surface in the days when many crofters took to sea for lobsters and other delicacies. Like the kelp industry and the bere meal mills they are now sadly long forgotten, and there are only two bigger boats left. 'In 1952 there was a terrible hurricane force wind. Then we used to have lots of hens here as we could be paid up to 4 shillings for a dozen eggs. But the wind was so violent that many of the hen houses were just blown away. It can be very savage here at times. I used to fish when I was a youth but I wouldn't go out now,' Bertie told me.

While the sheep have adapted perfectly to the shore, much of the rest of the island grows rich grass, and is the domain of large beef cattle. Yet more new regulations on transportation could make island life even more difficult as it has been suggested that livestock should only be taken off the island in winds of less than force 4/5. Since there are very few days in the year with less wind than this it could make life almost impossible. We stopped to admire Bertie's enormous stirks in a field close to his croft. Mountains of paperwork would have to be filled in before they could leave the island. 'In my lifetime it has changed so much, I really don't know where it will all end,' he said wistfully. Back at his cottage his collie was playing idly with a ball, and a couple of sheep were tethered close by on mowing duty. The sandy coloured flagstone roofs of the surrounding empty crofts were falling in, and the ubiquitous starlings had taken up residence. A fulmar was nesting up against an old chimney stuffed with marram grass. Bertie and his wife remember when all the crofts were occupied.

Leaving North Ronaldsay was extremely hard. With huge reluctance we boarded the plane. As she lifted agilely off the tiny runway on another tropically blue day, a huge party of seals was effortlessly flowing through the azure sea bordering the beach beneath us. Totally at one with the water, their sleek and graceful movement was sheer magic. As for the North Ronaldsay sheep, this supremely adapted animal will remain etched on my mind. Unlike many other breeds encountered on our Scottish journey, the very best place for this one is certainly their island home, the most glorious and fascinating I have ever visited.

CHAPTER FOURTEEN

Castlemilk Moorit

THE extremely rare Castlemilk Moorit is also probably one of the least recognisable of all the British sheep, but it is a fascinating animal, steeped in controversy and history. It owes its survival to Mr Joe Henson, who bought six ewes and a ram in 1970 for his Cotswold Farm Park. The sheep were advertised in the *Farmers Weekly* as 'Moorit Shetland Sheep', and came from the Castlemilk Estate, near Lockerbie, Dumfriesshire. The word 'moorit' means brown in colour, and is frequently used to describe the particular colouration found in primitive sheep breeds.

When Joe Henson first saw his new sheep, he was greatly surprised by their tremendous beauty. Despite the advertisement, they did not in fact resemble Shetland sheep at all, and for a time mystery surrounded them, for no one was really sure of their origins. Mr Henson regretted not buying more of these unique animals, but felt that Lockerbie was too far to venture to buy more. He was convinced that being so attractive, and unusual, the remainder of the flock would have been quickly sold, and he would therefore easily be able to find himself new rams at a later date. Sadly, on the first night, one of his precious new ewes was driven into a fence by a half-trained collie puppy, which had also come from Lockerbie. The ewe broke her neck, leaving him with only five. Devastatingly, he was later to discover that only another four ewes had actually found new homes, leaving the remainder of the flock, numbering more than 80 in total, to be slaughtered.

The animals had strong characteristics of the wild Mouflon, and of Manx Loghtan sheep from the Isle of Man. There were also similarities with the Soay, although these were later attributed to an earlier introduction of Mouflon blood. Over the years, records from the Castlemilk Estate showed that Sir Jock Buchanan-Jardine had bred his uniform flock of sheep from moorit coloured Shetland ewes, crossed with Manx Loghtans and Mouflon rams, possibly acquired from the mountainous regions of

Corsica. The result was a beautifully coloured, brown sheep with Mouflon patterned markings, and typical paler coloured bellies. The rams had strong thick horns, spiralling away from the cheeks, while the ewes' wide spreading horns gave them an alert and active presence. Wild and independent, they were bred by Sir Jock as park sheep, to join many of the other brown farm animals that he favoured. These included Guernsey and Ayrshire cows and, eventually, rare, dun-coloured Galloway cattle, which he was also responsible for breeding. His passion for brown coloured creatures led to the breeding of the famous tan coloured Dumfriesshire Foxhounds. After Sir Jock's death, the sad dispersal of the unusual moorit sheep was probably the result of the arrival of more productive, larger, commercial sheep.

Despite the fact that the Castlemilk Moorit was only established as recently as the 1930s, and was not recognised by the Rare Breeds Survival Trust until 1984, it is classified as a primitive breed having all the typical characteristics. Today, Peter King, formerly of the Rare Breeds Survival Trust, describes his flock of Castlemilk Moorits as having many outstanding qualities which include wonderful looks, delicious lean meat with a slightly gamey flavour, and superb wool. The latter is much favoured by hand spinners as it has little or no kempiness and, once spun, is extremely durable. Sir Jock had the wool from his early flock woven into cloth for his ghillies and keepers on the estate. In many cases the cloth was reputed to have outlasted the wearers.

It may seem odd that a breed of animals that started off its precarious climb from extinction numbering a mere five ewes and one tup, should have emerged with so few known health problems. Joe Henson's daughter, Libby, is an animal geneticist and together they have spent much time examining the Castlemilk Moorit. This is of particular interest in comparison to other breeds that do have health problems, despite a far greater gene pool. The animals appear to have no undesirable recessive genes, and consistently breed true to type, like peas in a pod. There have never been any abnormalities in the Castlemilk Moorit. This is particularly interesting when a comparison is made with other breeds with known health problems, and that despite having a far greater gene pool. Joe Henson was the founder chairman of the Rare Breeds Survival Trust, and it is very apparent, when speaking to him, that his passion for Britain's rare breeds is all consuming. Certainly, without his efforts this remarkable little sheep would have vanished into the mists of time.

On a glorious Indian summer day in mid-September we arrived at Glendy Mill near Glenfarg, Perthshire, where Malcolm Curtis keeps his small flock of Castlemilk Moorits. 'I have always been a bit of a back-garden farmer and bought my first sheep, Dorset Cluns, in the 1950s. These were followed by Jacobs, then endangered, in 1973, and by Castlemilks in 1994. They were then, and still are, at the head of the endangered species list – Category 1 (Critical), but I also had a personal interest as they originated from an estate not far from where my forbears were millers and farmers, near Dumfries – a double interest to support their survival. Delightful to look at, I admire them for their independence, for, like most of the primitive breeds, they have little respect for sheep dogs. Their approach is essentially a case of divide and confuse, "you go this way and I'll go that way, and we'll see what happens", but you learn how their minds work. They are also quite fearless.'

They are incredibly strong for their size, with streetwise fighting genes no doubt inherited from the Corsican side. Malcolm related the tale of a confrontation between a young resident Castlemilk ram and a visiting Jacob ram, a burly and mature chap with an awesome set of horns. At the first engagement, the young Castlemilk ducked under his challenger's horns to deliver a smart butt with his own. This was repeated and at the third clash the big Jacob was literally upended, a humiliation from which he never really recovered.

In the low September sun, the fleeces of the sheep were back-lit showing blonde highlights that many a lady would have paid a great deal for in a smart hair salon. The warm atmosphere was filled with cotton fluff seeds lazily drifting from the banks of rosebay willowherb on the verges. The alertness of the animals was most apparent, fine heads carried high and proud, with beautiful buff, brown, and beige hues mingling on their neat bodies. The rams were dozing by the gate, but leapt up when we went in to have a closer look. Once we were aware of their ancestry, it was not hard to imagine their forbears scrambling over rugged rocks amid herby, aromatic plants and shrubs in a dry Corsican landscape, while goat and cow bells tinkled in the valleys far below. The tups' horns were worthy of note: whorled and curled, gnarled and patterned as they circled round their cervine faces. Though most Castlemilk Moorits are now found on English soil, more small flocks are returning to their native Scotland. Seeing Malcolm's sheep in the magnificent setting of his converted seventeenth-century corn mill, grazing next to his field of inquisitive llamas, their future seemed very secure. But the dreadful thought of their brush with extinction was uppermost in my mind.

Sadly, there are no longer Castlemilks on the estate in Dumfriesshire. However, Malcolm is pleased that one of Sir Jock Buchanan-Jardine's relations has now started a small flock himself. Encouragingly, this is the third new flock to be developed in Perthshire, originating from the 1994 flock at Glendy Mill. With numbers well below 1000, it is important to preserve the breed as it has evolved, keeping it as true to type as possible.

'I think that the lambs of the Castlemilk are definitely the most attractive of all sheep breeds,' said Malcolm. 'Fleet of foot and like small deer fawns with beautiful brown curled fleeces, they captivate me totally. The ewes are the most wonderful mothers, very protective towards their lambs. They are a very hardy breed with little or no health problems, and are most attractive to look at.' As I took one last look over the gate, I found that the Castlemilk Moorit had captivated me too.

CHAPTER FIFTEEN

Achnacloich – Home of Scotland's Oldest Fold of Highland Cattle

THE bridge over the sea to the Isle of Skye has altered the atmosphere of one of Scotland's most romantic islands. However, the small ferry from Glenelg to Kylerhea still operates during summer time, taking cars across the swirling tide race of the Kylerhea Narrows.

For several years my family had the Old Inn at Kylerhea where we spent many holidays. Sitting on the rocks, watching the constantly changing currents racing in spirals like oily black patterns, it was hard to imagine that this was once the route for huge herds of Highland cattle leaving the island. These brave animals first had to traverse the perilous waters of the Sound of Sleat, before they even embarked on their long arduous battle through some of Scotland's most rugged terrain, on their way to the famous Cattle Trysts at Crieff or Falkirk. On many an occasion we watched inexperienced boat skippers turning back when they had timed their voyage through the narrows badly, and the tide race overcame them. It was always intriguing, visualising how frightened cattle would have fared in such forceful, treacherous waters. Many writers and travellers described the great cruelty to the livestock as they were roped together and forced to swim for their lives goaded on by the drovers in small craft alongside them. The Kylerhea crossing had to be attempted at hightide when the currents had eased a little.

For centuries Highland Cattle have been referred to as 'Kyloe' cattle, possibly due to the fact that so many swam across dangerous stretches of water or 'kyles'. It was not until the end of the nineteenth century that this barbaric practice ended and cattle were ferried in boats, in itself a highly dangerous operation. Often, drovers were injured or fell overboard into the icy waters, while their cattle badly gored one another with their horns. The rough boats were lined with scrubby branches to help stop the animals from slipping, and to try to absorb some of the vast amounts of dung they produced in their fear. On arrival at their destination, where there were no provisions for landing, the beasts were once more forced to swim. However, as A. R. B. Haldane suggests in his fascinating book, *The Drove Roads of Scotland*, 'the salt water was probably good for their wounds and helped to clean them up after a journey closely packed together.'

At one time there were hundreds of thousands of Highland cattle, spread throughout the Highlands and Islands of Scotland. They were the mainstay of the farmers and crofters, and were the only valuable asset that many of them owned. There was a vast traffic of animals to and from the islands and more remote parts of the mainland. Most of these beasts ended up being sold at fairs or 'trysts' to which buyers came from all over Scotland and England. Scotland has a long and bloody past in which cattle rustling was commonplace, for many animals and their drovers were killed in skirmishes and fierce ownership battles.

The Highland cow has witnessed many changes through the centuries. It was once more widespread than any other breed. Today, despite huge problems faced by many of our native farm animals, the Highlander holds its own and has once again become popular.

We first met Mrs Jane Nelson of Achnacloich at the annual Show and Sale of Highland Cattle at Oban. A familiar face at this twice-yearly event, she has scarcely missed a sale in more than 60 years. 'The sales always tend to go to a pattern and there is a great atmosphere. One very wet year, the mart was totally under water. After the war there was a big boom and the prices rocketed. Before BSE and Foot and Mouth prices were huge and there was a large demand for Highlanders from abroad. People at home wanted them too, but they just couldn't afford the big prices. Now things seem to have stabilised', Mrs Nelson explains.

Now in her eighties, she is a quietly spoken person with a lovely sense of humour, and is revered and loved by all those who know her. Her Achnacloich Fold of pedigree Highland Cattle was founded in 1901 by T. A. Nelson, the father of her late husband, Ernest. It is reputed to be the breed's oldest fold.

'They are wonderful cattle, they belong to the land to such an extent because they fit in and the land comes first. Highland cattle look after it.' Achnacloich is situated on the shores of Loch Etive, near Connel, and more than half of the farm is classified as a Site of Special Scientific Interest (SSSI). 'It is one of the most heavenly places, and I am so lucky to live here,' said Mrs Nelson who first came to Achnacloich in 1935, just before war broke out. 'I think my husband married me because I looked rather like one of his beloved Highland heifers, with my unruly red hair falling over my face. I arrived with a dowry of Jersey cows, rather like an African tribeswoman. The Jerseys used to boss the poor Highlands about; they are terribly bossy cows,' laughed Mrs Nelson, whose family lived at Invergarry. At that time, Highlands were crossed with Jerseys to produce a good hardy, milky cow for the crofters. Mrs Nelson's father studied the Jerseys' pedigrees before he bought them to ensure that his animals were extremely good milkers. 'We never showed the animals; they were bought entirely for their milking ability.'

At its peak, the Achnacloich Fold had about 80 cows. The fold is highly noted for its females, although there have also been many excellent bulls over the years. Perhaps one of the most successful bulls was Calum-Seoladair of Smaull, or Calum the Sailor, who was reputed to have been christened after an incident in the sea off Islay. He was bought for 100 guineas at the Oban Sale and, though not in the prizes that day, two of his offspring were sold for top breed prices at the time. He lived for 17 years, and was extremely active and healthy all his life, leaving excellent progeny.

The animals at Achnacloich are of the smaller traditional type. This is due to the farm's poor ground. Achnacloich means Field of Stones, which gives a good indication of the type of terrain. However, the smaller cattle still have very good growth potential when taken on to richer land. Despite the exceedingly wet winters, they are never brought inside except prior to a show or sale, when they have to be washed and blow-dried, and kept pristine. The occasional cow that has had a difficult calving may also require a few days' housing.

'We have always had the most wonderful farm managers here,' explained Mrs Nelson. 'Angus MacGillivary died some time ago. He was incredible because he had all our animals' pedigrees in his head and pedigrees from everyone else's too. A job was made for him here because he and his family knew so much about the breed and had worked with them for so long. The MacGillivarys come from the Isle of Mull and always rise to the top in everything they do.'

Now there are approximately 55 cows on the farm. They looked spectacular beside the waters of Loch Etive and its magnificent views. Mrs Nelson took us up a long hill track frequently crossed by racing, swollen hill burns. We passed the steading where contented cows and followers stood peacefully grazing in the soft incessant rain of the West Coast. A red squirrel darted up a gnarled pine trunk. The stunted oak trees were adorned with immense feathery lichens and the twisty birches appeared deep purple in the grey winter light. Such fine woodland would be dramatic in any weather. 'The cattle are put all over the farm at the appropriate time of year. They will eat everything in front of their noses: rushes, scrub and harsh grasses. The hill has improved beyond all recognition, thanks to their good grazing. Without them the farm would not be able to support our large flock of Blackface sheep. It's so beautiful up the hill, there are snipe in the bogs and my husband once found a nest of wild cat kittens up a birch tree. He was a very agile climber, and before we were married went out to St. Kilda with Lord Bute who owned the island at the time. Their aim was to gather some Soay sheep and collect enough wool from them to make a length of tweed for King George V. The wild Soays were almost impossible to catch, but due largely to my husband's agility, they managed to achieve their goal.'

We passed some lonely wild hill lochs; 'These are the three Black Lochs on the western border of the farm and are one of my favourite places. They have discovered some very rare dragonflies and medicinal leeches here too, as well as whooper swans, otters, and tiny brown trout. If you see a terribly dull, boring-looking dragonfly, it's probably one of the rare ones. Insects from the Arctic and the Mediterranean regions seem to meet up here.' Mrs Nelson explained.

'There are Achnacloich cattle all over the world and we have had buyers from Holland, Germany,

Original pencil sketch for finished painting

(Opposite) Achnacloich bull

Oak Woods, Argyll,

USA, Austria, Canada, Australia, New Zealand, Denmark, South Africa, and the Peruvian Andes, where a ranch of pure Highlanders was established at 12,000 ft. One bull we sold to Australia was an expert at breaking-out, and apparently left his offspring all over the district. A farmer in Sussex had some very wild Highlanders and rang to see whether we would sell him a quiet older cow. We sent an elderly cow of nineteen years old down to him. Not only did she produce several more calves, but she also calmed the others down. Normally we don't sell our old cows, that's the worst part. People still come here from all over the world to stay with me and see the cattle. I love to see them, I have met such interesting people and made so many good friends.

'You can always tell a good animal by the way it walks. Carriage is just so important. It's something you cannot describe, and so often you just do not see what you are looking for. The frame is very important, and of course the head. There should be a fine set of horns. These should not be turned down, and the crown should be nice and broad. I don't like the cows to be too long in the face. All our animals have tended to have excellent temperaments and we have never needed to use rings in the noses of our bulls. We used to show at the Royal Highland, Lorne, and Dalmally. We once took two heifers to Doune and Dunblane. One of them was champion while the other was reserve; it was very embarrassing when we appeared and swept the board. We were not very popular.'

Highland cattle can be a wide range of colours, from the deep, rich glowing red much favoured by German breeders, to light red, yellow, white, black, dun, and brindle. It seems that colours can be dictated by the fashion of the moment. Sometimes unusual coloured animals make higher prices at the sales. Most of these lovely colours can be found in the Achnacloich animals.

The cows usually start calving at the beginning of January and go on until early May. 'I was once given a present of some Highland bull's semen by the Duke of Montrose. It was rather a good present, don't you think?' Mrs Nelson giggled. 'That is the only time we have ever had any experience of using artificial insemination. We tend to feel that AI is inappropriate for Highlanders. Over the years we have crossed some of the cows with a White Bred Shorthorn bull, and now have a Beef Shorthorn that we use on a few of the cows. It was very hard to find a nice Shorthorn bull as the breed is beginning to die out.'

Mrs Nelson took us to see Achnacloich's fine garden that is open to the public in summer. Huge trees, including a dramatic Douglas fir, a vast Spanish Chestnut and beautiful Scots pines, mingled with azaleas, rhododendrons, and shrubs from Chile and New Zealand. The resident population of mallard has settled amongst the bamboo and water lilies on the pond, and unusual Lavender Aracauna bantams scratch in the mossy leaf litter. The garden slopes down to a point from where panoramic views of Loch Etive and the surrounding hills can be seen. 'Some evenings it is just quite breathtaking and I suppose, even after all these years, I still haven't quite got used to it. There are whole rafts of eider duck, wigeon, and otters in the bay, and that wonderful light. I just love it here. I don't actually feed the cattle or anything, I do hope you have not been here under false pretences', Mrs Nelson added in her typically modest manner. Having spent precious time in her company, I left with the strong feeling that the 'Old Breed' would be all the poorer without such genuine ambassadors as Mrs Jane Nelson.

CHAPTER SIXTEEN

Ayrshire Cow

THE night is hot and humid, the bedroom window wide open. On its sill a young tawny owl perches calling noisily. It is 3.30am. As I turn on the light it dashes away into a world of black drizzle. At 4.00am I set off in the car on my way to deepest Ayrshire. The roads are deserted; on their surface deep pools of rainwater are illuminated by the lights. The eyes of red deer grazing close to the moorland verges glint and a party of stags crosses close to the front of the car, while owls flit across the floodlit stage. The windscreen wipers are working hard. August. One of the wettest summers on record, and it is still raining.

Once on the dual carriageway a constant flow of lorries sends up spray in clouds and worsens the poor visibility. Even at this unearthly hour, passing the continuous sprawl that is Glasgow, there is considerable movement of traffic. I pass the Robert Burns International Memorial in the village of Mauchline at 6.10am. I am now late for the milking.

Once this rich countryside had a dairy on almost every farm. Sadly, growing pressures of milk quotas, huge demands on cows to produce massive quantities, and intense competition has pushed many of Ayrshire's dairy farmers out of business. Keeping up with today's pace is almost impossible.

Robert Cunningham greets me in the wet yard of West Mossgiel Farm with a cheeky, 'What took you so long?' as I follow him into the milking parlour. About 10 cows are already plugged in and there is the mechanical swish of the suction mechanisms on the milking machines, and the smell of hot bovine breath mingling with cow muck and disinfectant. The cows themselves are several feet above us, raised up on a plinth that runs round the centre of the milking machines and their tanks. The scene may have lost a great deal of its romance as there is not a milk maid in sight, and to be frank, with so much rain, neither the cows nor Robert are feeling at their best. 'It really makes me want to greet. We've

(above) A traditional Ayrshire Cow, West Newton Snowflake, 1906

00417

been trying to cut silage for weeks now, but the fields are far wetter than they are in winter. On top of that the poor cows are depressed and stressed with so much rain, and many of them have got sair feet frae standing in a' that muck. We've been feeding them since June. Half their winter rations are gone already, and the milk yield is well doon.' He releases the black suction cups from a well-formed udder, hangs them back up by the machine then daubs the exposed teats with special antiseptic. Heat from cow's urine rises up in a steaming mist filling the parlour with its distinctive odour, and there is a slopping sound as another cow pat hits the deck.

From my position it is hard to see which cows are the pure Ayrshires, but after a few explanations from Robert while he removes more machines, different udder attributes and conformation points are explained to me and I begin to see what I should be looking for. By now the milk is swishing through the machines almost soothingly, and the first batch of cows are pushed on and out of the parlour while the next eagerly shove one another into position and quickly start eating their rations. According to Robert, it has been many, many weeks now since any of them had a sunshine breakfast. The weather has been atrocious, and it is still pouring.

A third batch of waiting cows peer down at me. Coats beaded with rain, they have beautiful heads with the loveliest patterns and shading round their huge eyes. The Ayrshire is a very pretty cow, and varies greatly in her markings. While some are largely white with red leg markings, others can be almost pure red, or mahogany in colour with markings like eye-shadow on their faces. Many have patterns resembling dark red maps that dot an ocean of white on the rest of the body. Black and white too is occasionally seen, and is still referred to as the Ramsay Ayrshire. Though their legs are caked with mud, they look healthy enough, but like their owner do seem to be depressed by the prevailing gloom. He washes the udders of each of them with constant running warm water from a long pipe before putting the machines on, and always puts some antiseptic on their teats after they have been relieved of their milk. Like page three girls, a dairy cow's milking tackle is her greatest asset and must be kept in good order, although for a cow, biggest does not necessarily mean best. The mud causes teats to split and crack like hacks on the fingers of a deep-sea fisherman, while the dreaded flies help the spread of mastitis which all too frequently can totally write off an animal's milking quarter. With little sun, the cows seem to have lowered fertility rates. And on top of that the muggy climate and constant wet is perfect for pneumonia. Robert Cunningham's vet bill will be exorbitant this summer.

The Ayrshire was first recognised as a breed in its own right in 1814, and the first Herd Book Society established in 1877. The early cattle were sometimes called the Dunlop, or appropriately, the Cunningham. The Cunninghams have been breeding top quality Ayrshire cows on their 200 acre Mauchline farm since Robert's father, Bobby, first arrived in 1949 and had a herd of 25. Then nearly every farm in the area kept traditional Ayrshire cows. Today they have approximately 100, and although most of the herd is pure, the red Holstein, a large angular, high yielding beast is fast filling their place. This cross may still be registered with the Ayrshire Cattle Society provided that the percentage of Holstein blood is clear in the pedigree, and that they are not shown in classes for pure-bred animals.

Bobby appears at the entrance to the parlour with a large empty bucket. This is filled directly from the tanks and he disappears off into the murk to feed the calves. As I catch up with him, he explains that the youngest calves are fed on fresh milk to begin with, and then are put on to a powdered milk supplement. In a large byre individual pens of calves await their breakfast. There is the sound of sucking as they butt and slobber on their neighbour's faces in keenness for their breakfast. The little buckets of the youngest calves are filled with the newly drawn milk and in seconds it disappears and they return to sucking one another's noses, cleaning up every last trace of milk in the process. Then Bobby measures the powdered milk in buckets assisted by a small, persistent black cat that leans right into the mixture while snowed with milk powder, and laps greedily. 'The cats are vital here and keep us clear of rats so we don't ever begrudge them as much milk as they want,' Bobby says. Hours after their birth,

the calves are taken away from their mothers. All the pure-bred heifer calves are kept, while the bull calves go into a beef production regime. The cows are then milked for about 10 months and given a two-month rest before the arrival of their next calf, which once more starts off the whole procedure. The heifers are always put to a pure Ayrshire bull, though some of the older cows go to a Holstein or Holstein cross Ayrshire bull, or have AI. Calving goes on all year round.

Armed with empty buckets we set off across the mire in search of concentrates and bruised barley. 'We have found that it is cheaper to buy in barley, and this year we are trying to make whole crop silage with the barley grown on the farm. But we now need several weeks of dry weather – one day is no use to us at all. I never remember such a terrible summer.' The air fills with thick dust from the bruiser as Bobby fills buckets and sacks. It is clear that the strain is beginning to show. As millions of plastic containers of milk are poured over breakfast cereals all over Scotland probably as we speak, perhaps few consumers imagine the hard graft that goes into its production.

Another black cat snoozes on top of some scales in a byre as we pass through armed with feeding. 'I used to be fair proud of this byre. When it was first made it was really modern and up to the minute. The coos were inside in the winter in these stalls. Then they all had horns and we used to have to put weights on to them using a pulley system to train them to have a nice curve, sweeping forwards then back.' On some farms horseshoes made ideal weights, but care had to be taken not to be knocked out by them. If the cow bent to eat at the wrong moment then the stockmen were clouted on the head by mistake. Once the cows went into loose housing, horns became more of a hazard as they could injure one another with them so they were removed. Today, calves are disbudded at approximately 10 days old.

'My father always had Ayrshires too, and he used to help our neighbours preparing bulls, washing them in readiness for the sales. In 1949 the breed record of £9000 was achieved. There used to be a week in November when there was an Ayrshire Sale every day. They were so popular at that time. Lots of Ayrshires went out to Ireland, and still do. But now Finland has the largest population. Of course I would like to see all pure Ayrshires on our farm as they are unbeatable dairy coos.'

The reputation of a dairy bull has always been very dubious. Unlike beef breeds they are renowned for their fiery, erratic temperament, and must therefore be handled with the utmost care. In a large pen next to the calves a huge bull with very small eyes chews his way through a pile of grass clippings. He watches me with an untrustworthy sideways glance. Bobby puts a large bar behind him to keep him in the small division of the pen so that we may safely muck him out. 'We take no chances with him. He is only here for a short working holiday, and he does not have a nice temper. It is always the quiet ones though that are the worst as you are usually ready for the ones with the reputations.'

On the other hand the reputation of the Ayrshire cow is excellent. She has a placid, kind nature, great longevity, deliciously flavoured milk high in butterfat, and is extremely hardy. One of the only reasons why Robert now crosses some of his cows is financial pressure. 'A while ago a cow was thought to be doing well if she produced an average of 1000 gallons of milk a year. Now she is expected to produce a 2000-gallon average, and the only way she can do this is to have some Holstein blood added. The old fashioned Ayrshire coo was far too small to keep up with today's pressures. We will always keep a percentage of pure Ayrshires, but we do need some of the others to keep up with the market. The performance of every cow is closely monitored and that is what counts. Milk records are really vital, because when you sell a beast, it is her records that the buyer will be looking at,' Robert explains. 'My father would like to see all pure Ayrshires on the farm, but unfortunately, I just have to move with the times, and putting milk into the tank is what is important now.'

We retire to the house for breakfast where I witness that fresh Ayrshire milk straight from the cow does taste very different, and is far superior to the milk we drink daily. The Cunningham's milk is collected after the afternoon milking and goes away to the Lockerbie Cheese Company. Once this part of Scotland was famed for its butter and cheese, and most Ayrshire milk went to this industry. A new scheme, White and Wild, aims to pay a premium for milk produced on farms where they retain hedgerows and wildflower headlands, something Robert approves of wholeheartedly.

'The Ayrshire used to be made up of two different lines. Those that were bred particularly for their udder conformation, and those that had the looks. The breeders of the two types were referred to as the *milk vessel boys*, and the *yeld stock boys*. The shape of the udder was most important to the vessel boys, and unlike many commercial dairy breeds, vessel-bred cows had very neat, yet capacious udders with a great length, and perfect teat placement.' The teats of an Ayrshire cow have always been small. Traditionally the cows were milked by women, and small teats proved easier to handle. 'Teats should be pointing straight at the ground. East and west simply will not do,' laughs Bobby. 'The yeld stock boys concentrated more on breeding coos for the show ring with a flashy appearance. They were big, showy eye-catchers. What I like to see now is a big coo when she is lying doon, a large engine room to produce all this milk. The modern Ayrshire is now much bigger. There is no doubt though that she will go on far longer than the Holstein. Many of ours go on for 12–13 lactations – the Holstein simply can't do that. But we feel that we are really doing well if we get 10 lactations out of any coo,' Robert explains.

For the Scottish dairy farmer, the main show to win is DairyScot at Ingliston, in November. 'Cow of the Year Award' is the ultimate goal. In a moment of idle thought, I remember how we bestowed this title on to some of the fiendish female bullies who made early school days sheer hell. In their case a prize effortlessly and unknowingly won.

Ayrshires always had a good straight top line across their backs, and at one time were only clipped round their head, tail and udder. Clipping round the udder is important as it makes the milk veins more prominent and shows a cow's capabilities, proof of a potentially good milker. Now the whole cow is clipped before a show, except along its top line. For the big shows, the Holstein breeders bring in fitters, and this trend is fast spreading into the show ring for pure Ayrshires too. These professional Bovine Beauticians come dressed in immaculate uniform overalls, armed with a bag of hair dryers and suitable cosmetic products for the discerning cow. In some case the fitters visit the farm six months before the show to carry out an initial clip and discuss the dietary requirements of individual beasts. So it seems that it is not only the stars that have personal trainers. However, Robert only uses the expertise of these skilled Cow Coiffeurs to finish off his cows' top lines, making them appear beautifully straight and eye-catching. Before a cow goes to the show, Robert gives her half a bottle of neat whisky. 'This has always been done and they say it helps to keep their stomachs right and keep the milk in the right place. My father once gave his cow 5 bottles, and she was reputed to have staggered her way round the ring. I think that was a bit much.' He laughs. Following the cancellation of shows during the Foot and Mouth outbreak in 2001, Robert has not been showing this year. 'It's very expensive and at the moment there are far too many more important things to do here. We will show again, but the new regulations just make things a great deal more complicated. My father loves showing anything, in fact he would even show a donkey.'

While beautifying a show cow has now become an almost scientific art form, bovine photography is a painstaking process that involves no less than 5 people. One holds the subject's head, while a second rattles a bucket of feed to keep her attention. A third raises the front feet up on to a plinth, while a fourth is in charge of the hind legs and tail carriage, aided by an almost imperceptible piece of fine thread that holds the tail in perfect pose. The fifth person is the photographer. Meanwhile, the subject has to stand impeccably, usually in front of a painted canvas backdrop. This is worse than being a supermodel photographed for the front cover of *Vogue*. The end result looks fabulously professional, and even if you did not successfully become Cow of the Year, the portrait probably makes your owner feel pretty proud of you, particularly when it is accompanied by an impressive set of milk records.

The rain has cleared a little leaving fine drizzle like hairspray. We set off past Mossgiel, the neighbouring farm where Robbie Burns farmed between 1784 and 1786. A gnarled and knotty elm tree trunk has developed a growth like a face reputed to be that of Tam O'Shanter, one of his most famous characters. Now the cows are cudding quietly and watch us as we wade through the muddied gateways, our boots oozing and sucking noisily. Gentle beasts, they seem resigned to the climate and many lie dozing on grass laced with beaded cobwebs. Though the continual deluges have been ill-suited to farming, the hedgerows are thriving and look thick and lush in the landscape dotted with the yellow perilous ragwort. Despite today's strong Holstein influences on the dairy industry, there is no doubt that the Ayrshire will always have her place. With a huge increase in the organic market, she is perfectly suited for this system with an ideal temperament, excellent longevity and hardiness. Like much of Scotland's native farm livestock, her ability to survive on poorer ground too, is a vital attribute not to be overlooked. The Ayrshire is the ultimate, economic dairy cow.

The sun peeks through a particularly dirty black cloud, and by the time I have passed Glasgow, is holding her own once more. Home by mid-afternoon, the day seems still young, and Ayrshire a long way away. By the time I am on my second cup of tea (with milk) hard-grafting Robert will be starting the afternoon milking. The tiring life of a dairy farmer is unrelenting. I hope that the sun will bide long enough to raise his spirits and allow him to salvage their sodden silage crop, for as he says himself, 'Though it fair gets me doon at times as it has become so pressured, dairy farming is all I know.'

005

CHAPTER SEVENTEEN

The Northie

LAIRG, in Sutherland, in the far north of Scotland, is a very long haul from most places. However, during late summer and early autumn, a large group of enthusiastic and loyal pilgrims wend their way to this small unspoilt location lying north of the Dornoch Firth, at the bottom of Loch Shin. For most of the year the quiet town remains unremarkable and its auction mart is silently dormant, the old wooden sheep pens empty, its sale ring and building forlornly abandoned except by spiders and mice.

Yet from the middle of August it comes alive, bursting forth with activity as it is transformed into the nerve centre of the North Country Cheviot. The pens are checked and numbered, the rings cleaned, and the floors covered with sawdust in readiness for a huge ovine invasion. For Lairg Auction Market holds the largest one-day lamb sales in Europe, to which stockmen bring their sheep from all over the vast tracts of wild country further north, as well as many hundreds who travel from the south.

If you have never been to an auction mart, any one of the four sales held here is quite an inauguration, and leaves a lasting impression, for despite the continual changes in agriculture they remain unique. The market itself is largely un-modernised, and it is its atmosphere that is so extraordinary. It buzzes with excitement, particularly on days when as many as 40,000 lambs go through the ring. In its heyday, before new rules and regulations, livestock health scares, and the dreaded paper-mountain that has become an integral part of the livestock farmer's work, this was one of the most infamous events on the Scottish Sheep Farming Calendar.

As you approach Lairg from the south on a sale day, the small road on the edge of the town is lined with a variety of livestock vehicles, lorries and cars for a considerable distance before the mart's old building has even come into view. Buyers from Cornwall to Caithness descend here, not only bringing

(above) Mrs Mackenzie

vital business for the sheep industry, but also for many small local businesses, such as pubs, shops and bed and breakfast premises, particularly important in such a remote rural community. Before livestock haulage was readily available, the sheep were walked from outlying farms and crofts. Even after a long and strenuous trek of 8–10 miles a day, they usually ended up at their destination in a fatter condition having feasted on lush grass en-route.

Lairg's autumn sale of ewes and tups was Keith's maiden visit to the world of the livestock market, and it did indeed leave a great impression on him. He was enthralled. It was one of the first places we went to as we began our long Scottish farm animal tour. As we walked into the packed sale ring, the auctioneer, David Leggat, was in the throes of selling a tup. Despite this, his eagle eye caught sight of us and he greeted us, cleverly continuing his bidding at the same time. This unique vocal skill is very much a part of his personal auctioning technique, humour and charm. The atmosphere was electric and the staggered seating boards in the ring's amphitheatre were crammed with bodies, the smell of tobacco smoke mixing with the aroma of sheep, sawdust, hotdogs and bacon rolls.

A familiar, much-loved and admired member of the British livestock-farming scene, David started out as a trainee auctioneer in the offices of United Auctions in 1975. Here he licked a great many stamps and, as he says himself, became very good at it too. From a west of Scotland farming background, he has always had a passionate love for any good quality livestock, and is a particular supporter of native breeds. He began selling sheep a couple of years later and soon became an accomplished and popular auctioneer. Today, he is a director of the United Auctions Group and represents their livestock interests. However, he is also still heavily involved in the auctioning side of the business and loves his visits to Lairg every year where he does much of the selling.

'The sales at Lairg were started in 1895 because Sutherland Estates had so many sheep in the area and needed somewhere to sell them. Lairg was the obvious centre being close to the railway and accessible from much of the north. They are a unique event, and are really one of Sutherland's biggest days. The wonderful thing about Lairg is that you will not only find consignors who have just one or two sheep to sell, but also crofters and farmers, and then the large estates that bring in huge numbers. Lairg is full of colourful characters and the sales are a real social event. The market bar is always packed and is the scene of much celebration or commiseration. In the past, some of the shepherds and indeed their employers too, ended up pretty pickled at the end of a long sale day. It has been said that the way they found their colleagues was by recognising all the different estate's tweed breeks. I have to be very careful you know, because of course if I am doing the selling, I cannot afford to be under the weather.' David laughs, but has been known to have a good spree himself every now and again.

'The North Country Cheviot has been in Sutherland for over 200 years. It is an amazing breed as it really does meet modern demands. Much of Sutherland and Caithness is hard, high, and inhospitable, but through the course of time, with the help of good breeding, natural selection and the true dedication of the flock-masters, they have ended up with a fantastic animal. Unlike other breeds, the North Country Cheviot tup is not usually sold as a shearling, but frequently taken to auction as a four-shear. In some areas they have a very disciplined line-breeding scheme whereby years after a particular animal has been sold, some of his offspring are bought back to keep up the line. I always feel that the Lairg men have their feet firmly on the ground. Though there can be some high prices paid for the tups, you will not see the fireworks and astronomical prices that occasionally happen in other breeds.

'It was terribly upsetting during the Foot and Mouth outbreak because the Lairg sales were cancelled altogether. It affected everyone, not least of all the small businesses in the area. This is why it is so vital for them to continue. Not only that, but being the first big hill-lamb sales of the season, they tend to set the price and pace, and act as a gathering, sorting, and dispersing point. This is so important for the sheep industry. Following the Foot and Mouth episode, there is pressure to alter

Ewes at Lairg Sale

Lairg, or stop it altogether because it is not like many modern markets and cannot be easily disinfected due to its wooden construction. It would be tragic to lose this event, and I really believe we must keep it going at all costs, for the social side too.' David is adamant.

Outside the main building a maze of pens is packed with Cheviots of many different types. Pristine and well prepared for their debut waltz round the main ring, they cud quietly as they are studied by a continuous throng of appraising eyes. Huge boots with studded soles and turned up toes scuff out a roll-up in the dust, and collies help to guide some late arrivals into their allotted pen. A couple of characters as rugged as the Sutherland hills, dressed in denim dungarees topped with patched tweed jackets are putting the world to rights in a corner, while a pipe-smoking John McNab in a tweed plus-four suit has clearly found the tup of his flock's desire. 'Oh, there you are Mrs Mackenzie, just like a War Memorial' exclaims a ruddy weather-beaten shepherd as he spies old Annie Mackenzie eyeing up a pen of tups in a sun-speckled corner near the entrance to the main ring. Ruairidh Mackenzie and his mother Annie have been familiar faces at the Lairg sales for as long as anyone can remember. Annie laughs at the comment, and explains that Lairg is a very social event and one that she would not dream of missing. 'Usually Ruairidh and I go to Lairg together, and when we arrive he says, "Mother, I am just off to have a wee word with so and so, but I'll be straight back." Of course I just know fine well not to believe a word of it as he loves to chat and always disappears for hours on end, but I don't really mind as I see all my friends too. People always used to come to Lairg for the crack even if they were not buying or selling sheep. Luckily, Ruairidh and I think the same way about sheep, but if we do disagree, I can always hit him as he can't retaliate,' she jokes mischievously. With a face lined and brimming with character, and usually wearing a hat of some description perched at a jaunty angle over her straight white hair, Mrs Mackenzie stands out in a crowd.

She grew up on a croft in Knockfarrel, near Dingwall, and has always lived for her animals. As a teenager she regularly walked the sheep from the croft to the station with her collie, Nell, and can recall helping to load approximately 40 into each railway carriage. 'There were no ear tags nor forms to fill in then, we just put the sheep on the train and off they went to a buyer further south,' Annie explains. Most marts were at the rail-head, and sheep as well as lambs straight off their mothers were frequently put on to the trains.

'A good stock man is someone who can tell a beast that will be sick tomorrow. This is a gift that few people have, but it certainly applies to my mother,' says Ruairidh. 'Mother is wonderful with stock, but she does not suffer fools gladly and even now that she is retired likes to keep a close eye on all the sheep. She has been involved with the Northie since she was a wee lass, and as soon as I was born took me off to the Lairg sales in my carrycot. Father was a great *kenner* too, by this I mean that he really knew each sheep individually. I will never forget one year there had been a freak liaison and one of the ewes, most unusually, had been tupped in August. Father, who was an MP, came home for the Christmas recess and was checking the sheep. He said that the ewe should have lambed so we ran all the sheep past to look for her, but he knew immediately that she was not amongst them. Reluctantly we all went out with Father to go and look for her, and after a great deal of searching found her behind a bush with a nice set of twins. I remember it clearly because it was during the afternoon of Christmas Day, and we were upset because we were desperate to get back to watch *Top of the Pops* on the television.' Ruairidh laughs.

The Mackenzie family have been tenants of Heathmount Farm, Tain, in Ross-shire since 1933. The farm consists of 600 acres of rough pasture, and approximately 200 acres of arable ground. They have 600 North Country Cheviot ewes and as Ruairidh says, they suit their ground perfectly. Ruairidh has two daughters, Kirsty and Mairi, and though keen on farming, he is not sure that they will be able to continue. 'Sadly, now it is not a good business to be in, and in order to keep the home fires burning I have to look elsewhere for an income and have several part-time commitments. I am a Director of the

British Wool Board which meets in Bradford in Yorkshire. There we try to obtain a good return for the producer, but it is very hard during the current agricultural climate. Britain still has the dearest wool in Europe largely due to the Wool Board's efforts. From the farm here our wool is first graded locally with that of other producers, and then is auctioned in Bradford and usually sold to buyers for carpets, and occasionally cloth. Then it will go on to a scouring plant. The Cheviot has very good quality wool. During the Foot and Mouth outbreak, collecting fleeces from anywhere in Britain was a big problem. Then all the buyers bought foreign wool and blended it with existing stocks of British wool. That certainly did not help the situation and our wool was worthless. We can remember when the wool cheque would pay the shepherd's yearly wages,' Ruairidh explains. He is also a Director of the Farmer's Board which promotes United Auctions in Northern Scotland. 'Basically, I am the man who takes all the flak,' he laughs, 'but I thoroughly enjoy my work as it is always interesting. It does not pay for me to work at home all the time but we have a very good shepherd, Willie Muir, who has been with us now for many years.'

'When I married Alasdair Mackenzie and came to Heathmount I was still in my early twenties. My husband was a truly great sheep man and always put the farm first, but because he was so public-spirited he also ended up being on lots of County Council committees. He finally became Liberal MP for Ross and Cromarty between 1964–1970. Due to this I ended up doing most of the sheep work with the help of the rest of the family, but I never minded. In fact, I thoroughly enjoyed myself,' Annie says. At this she takes out a yellowing cutting from a local newspaper that reads as follows:

> That chestnut about a business running more successfully in the absence of the boss was given another airing at the weekend when the Scottish Liberals were in conference in Inverness. Alasdair Mackenzie, MP for Ross and Cromarty and a sheep farmer from Tain, was not at home lambing with arguments on the Finance Bill and trotting along the lobbies of the House of Commons. Meanwhile his wife, Annie was running the farm. As a result of this his colleagues were teasing him that it had been the most successful lambing since he started farming in 1933. From 555 North Country Cheviot ewes there had been 347 pairs of twins.

'Well,' says Mrs Mackenzie, 'that just shows you doesn't it?' She proudly folds away the cutting with a very elfish smile and then brings out a photograph of herself in the *Scottish Farmer*'s Living Legend feature surrounded by North Country Cheviots. Mrs Mackenzie is well thought of in North Country Cheviot circles and is clearly every bit as hardy as her beloved flock of sheep.

The North Country Cheviot, fondly nicknamed the Northie, is the largest of all the hill breeds. As its name suggests, its origins are in the Cheviots. However, during the 1780s a well-known agricultural improver, Sir John Sinclair of Ulbster in Caithness, was concerned about the quality of many sheep flocks both in Scotland and England. He believed that they could be easily improved with careful selection and some new blood, and set up the British Wool Society at the time. He took a flock of 500 South Country Cheviots to his hill farms in Caithness. Here they thrived and proved to be highly successful, being better suited to the ground than the Blackfaces that had previously been there. As time went on, huge flocks of Cheviots were moved into crofting areas, as they were tragically vacated during the Highland Clearances when families were driven out in order to make way for them. Agriculture was altering drastically across much of Scotland at that time, and the Cheviot was an integral part of this change. Unlike today, the price of wool then was high, and in order to produce sheep with a better quality fleece, some blood from the wool-producing Merino was gradually introduced. This slightly changed the characteristics of the northern sheep eventually dividing the Cheviot into two completely separate breeds. In 1912 a North Country Cheviot Breeders Group was formed which enabled the new breed to be shown, and after the Second World War in 1945, the North Country Cheviot Society came into being.

The North Country Cheviot tup is occasionally horned, and a Roman nose is a typical feature. It is very popular and produces an excellent lean meat lamb when used as a terminal sire on many different hill breeds, including the Blackie. The Scotch Half-bred, occasionally referred to as the Queen

Tup, Heathmount Farm, Tain

Ewes and lambs, Heathmount

of Sheep, is a cross between a North Country Cheviot ewe and a Border Leicester tup. The large female progeny of this popular cross produces a high quality meat lamb when put to a Down tup such as a Suffolk, or a continental sire. In the Shetland Islands, the Shetland-Cheviot is another recognised half-breed still immensely popular today. The result of a Northie tup on a Shetland ewe, the progeny are extremely milky and hardy and make excellent breeding sheep that can then be put to a quality terminal sire. Economical to keep as they do not require a high feed input, they are ideal for Shetland's impoverished ground. The sight of small neat Shetland-Cheviot ewes being buffeted from underneath by huge suckling lambs much larger than themselves is a familiar one.

Like the Blackface, the North Country Cheviot has developed into three distinctive types: the Caithness type, which suits the better, more productive ground of the county and is a substantial large animal with a prominent head, while the smaller hill-type or Lairg Northie, suits the hard ground of Sutherland. A third animal, the Border Northie, developed in Scotland's Border region, during the period of time between the World Wars, is a large and prolific sheep. While the South Country Cheviot has alert, erect ears set on the top of its head, those of the North Country Cheviot are set at approximately 45 degrees, a throwback to the addition of Merino blood. This telltale feature makes confusion of the two types more difficult. 'The Southie is usually easy to distinguish from the Northie as they tend to be very consistent – little and wild,' Ruairidh jokes, 'But don't tell them that I said that.' In 1981 a Hill Sheep register was started specifically for breeders of the hill, or Lairg Northie.

The Mackenzies favour the Lairg type, and says Ruairidh, 'We like them because they really do suit our ground and it is important to match the stock to the land.' Many of their fields are surrounded by rough moorland that during our visit was a carpet of waving bog cotton. Gnarled and wind-blown ancient Scots pines bent in primeval shapes add sculptured form to the beauty of the surrounding scenery. Everywhere yellow gorse was in full flower. In the distance an osprey called tetchily. While the rest of Scotland had been intolerably wet, this lovely north-eastern corner on the edge of the Dornoch Firth was very dry and had somehow escaped much of the late spring's incessant rain.

A group of large lambs was playing round an old stump in one of the fields at Heathmount. Their tremendous body length, a vital breed characteristic, was clearly apparent as with great athleticism they sporadically raced one another up and down the field's edge, leaping the stump as they went. Under the right conditions, the Northie ewe is prolific and is renowned for her excellent mothering abilities and milkiness. As well as these attributes, she is also known for her longevity. Annie Mackenzie recalls a trip to Lairg in 1978 when a pen of their five-year old cast ewes made £76 each, and went south to be crossed with a Border Leicester sire, to produce several more crops of excellent half-bred lambs.

There is clearly something about the North Country Cheviot that arouses excitement. So much so that on one particular occasion while David Leggat was selling a tup at Lairg, he got so carried away as he was bringing his gavel down that he somehow let go of it and it shot up and whacked him hard between the eyes. 'It was bloody sore and for a moment my vision went skew-wiff and that tup came and went in the ring. I really thought I was going to black out. Luckily I recovered quite quickly. Unlike that of the Northie, my head is not very strong. I clearly remember on one occasion, a tup managed to bash his way out of the side of a Land Rover with his head on his way from the market. The new purchaser stopped to round him up again, but all he could see was his animal swimming away out into the Dornoch Firth. It was terrible, as sadly, that beast was simply never seen again. These are animals with great determination and strength, and I have nothing but admiration for them as they survive in the harshest of conditions.'

CHAPTER EIGHTEEN

On the Border

A FRANTIC whistling is coming from the distant fields accompanied by banshee wails of frustration. Almost with the speed of sound, a collie returns to the yard. In her wake, a creature with flared nostrils, snorting furiously and reminiscent of the devil incarnate, storms to the gate. A hill Blackie has got the upper hand again. Collies, like computers, are only as good as their operator. You may go out and buy yourself the best and most expensive model, but alas if you do not know how it works you will certainly not achieve much of a tune out of it. There's another fact too, that however good you and your dog are, bloody-minded sheep can always surprise you.

The Border Collie is yet another of my passions. Though I am filled with awe while watching the magnificent dogs and their handlers that win trials, and the brilliant contestants on 'One Man and His Dog', it is the good old-fashioned farmyard collie that fascinates me the most. Many of these stalwarts of the farming community are often much-maligned, whiling away their time wandering round the farm, rolling in unsavoury morsels and eating a bit of this and a bit of that while awaiting their next instruction. Then eagerly (frequently too eagerly) they will spring into action to do as they are bidden. That is the theory at least. Many under-worked dogs are far too keen, and when instructed set off with so much enthusiasm that they end up rounding up everything within a mile radius of the homestead. This often includes the neighbour's sheep, the pet goats, the farmyard fowls, and any stray children too, if they happen to be in the way. Other dogs go off misleadingly steadily, but when given their orders suddenly develop that infuriating ailment, IMD – intermittent malevolent deafness, a complaint also frequently found in old men who have had to put up with years of nagging. There is no cure. I have often been witness to the yelling and swearing that seems to accompany some dog handling, and must admit to my own share of shouting. But all the farm dogs we bought as trained were only inefficient because of us, and as we raised our voices, swore and lost our tempers all we succeeded in doing was confusing the issue further. A collie is only ever as good as its handler. But all too often this basic fact is overlooked, and the dog is shut away and tagged as useless, manic, stupid, or impossible to train. Bad workmen always did blame their tools.

As Jim Hogg of Ballechin Farm, Ballinluig, Perthshire says, 'Almost every poorly trained collie that does not work properly has its owner to blame. People are very inconsistent with their animals and resort to a good thrashing far too quickly. A severe voice and a good shake is all that is needed, and above all constant praise and reward when something is achieved. People forget all this and then wonder why they cannot get their dogs to do what they want. You really do have to go with a dog and try not to pick up on every little mistake', Jim laughs.

We stand outside his house on an evening of glorious sunshine with the distant hills dappled with fading light. Meanwhile ten collies are going berserk on top of a steaming dung midden, playing mad dog, chasing one another round and round. Some are plunging into a deep pool of thick, stinking effluent to cool off at the bottom. An eleventh dog clings to Jim's side. 'That's Vic, he came to me a few months ago having been a complete failure with terrible nervous troubles. It took him weeks to even let me touch him as he just cowered all the time.' He bends to stroke the tri-coloured dog and there is no sign of fear now. While the rest of the gang play, Vic is showing a little interest, but almost seems frightened to be seen enjoying himself. However, it is obvious that soon his curiosity will get the better of him.

'Och, just look at those dogs, they are filthy. I shan't be wanting them on the settee tonight, but it is important for them to play as well as work', Jim laughs. 'That reminds me of one old bitch I had who always liked to secure a good place on the settee but was usually pushed off by the younger, fitter dogs. They used to all pile into the house and there she was ousted as usual. So she would quickly go outside and start barking as if there was someone arriving in the yard. Immediately all the others leapt off and belted outside to see who it was, and like lightning she was back in there, right on to the middle of the settee. Though she did this time and time again, they always fell for it. It was dead funny. These are dogs with an incredible brain. That's one of the many reasons why people go wrong with them as they do not give them credit for such intelligence.' He whistles up the muddy pack and they return obediently from their antics vying for his attention, all except one young pup that disappears off to his tattie patch, returning shortly crunching up a potato. 'You can teach a collie to do almost anything. I know of one man up in Orkney who taught his dog to fetch things for him, including peat for the fire and food from the cupboards, and to open and shut doors. They are easily taught. I used to be able to do Scottish dancing with a collie I had called Glen.'

Dogs are in the blood of this family. Jim's father was a great dog man, and his brother Angus, head keeper on Remony Estate, on the side of Loch Tay, has a reputation for his gun-dog training skills. Jim, who has been a shepherd all his life, has built up a considerable reputation for sorting dogs out, and receives so-called 'failures' from all over the country. 'People used to ask me to help them and I would work away getting the dog going properly, then once things were going well, they would just take it back, and no doubt the rot quickly set in again. Perhaps I got a bottle of whisky at the end of my efforts. Now I don't do this anymore, but when someone rings with a problem dog, I buy it from them for the pup price, which works out much better for both me and the dog. Eventually I will sell it on to a suitable owner. I have really only had a couple of dogs that I couldn't do anything with, and one of them definitely was stupid, but it's very unusual. I have a friend who bought a middle-aged collie from Battersea Dog's Home and we soon got him working. It is in them all to work. As I said before, it's the folks and not the dogs who fail almost without exception. I can aye make something out of a dog, but not a handler.

'When training a collie it is the tone of voice that is most important, and as it starts working you

have to move yourself back and forwards to accommodate its movements. We are the ones who are up on our feet, so we should move to suit the dog, not the other way round. You do have to be very firm at times, and that's where the bunnet comes in very handy. When my bunnet's off and the dog sees me doing three strides to the acre, it knows that it has to behave itself. I always have a deerstalker and the dogs often chew it. The flaps on it are very useful as they stop the brain from blowing off. Before the arrival of the quad bike, farmers and shepherds walked miles and miles checking sheep. This calmed a dog down. But nowadays many of them are far too fresh when they start working. This really doesn't help the situation, as they are just too inclined to go off like a rocket. I can't be bothered with quads, its far too much hassle getting on and off to open gates when I can just throw my leg over the fence instead.'

Unlike Jim, many modern farm collies travel around with their owners on a quad bike. They have learnt how to lean into the wind on the side of a steep brae face, ears flapping in the slipstream, stiff legs keeping balance as the bike bumps over the roughest terrain. They can perch precariously almost anywhere on a tractor, clamber into the bogie on top of soil, dung, wool, rubbish, tools, or the unfortunate bodies of the fallen, or travel on top of the livestock farmer's paraphernalia in the back of the pick-up or Land Rover. Collies aren't fussy, any transport will do. While they can vary in temperament and type enormously, most of them have several things in common: natural inquisitiveness, a great desire to please, and an even greater one to join in.

All the dogs that we bought as trained were always very willing and obedient, their short-comings purely a result of our erratic instruction. But some of the sheep on the farm were little short of hell when it came to faultless displays of dog handling skills. We used to buy cast Blackface ewes from Colonel Mavor, Braes of Foss, a farm situated on the exposed side of Schiehallion, one of Perthshire's most famous landmarks. They were the wiliest sheep that I have ever come across, yet always did well with us providing we could put up with their untameable natures. When crossed with a terminal sire most of them would effortlessly rear a good pair of twins, but we always dreaded encountering any lambing or health problems as they were inclined to go into orbit with the least provocation. They were sheep of great fortitude, and even during the hardest winters a few refused to come to the trough for feed, preferring to keep their independence even if it meant ending up looking as thin as a supermodel. When we tried to shed off any particular ewe, she faced up the dog with incredible defiance. One devilish old girl with windswept horns learnt to drive the poor dogs straight at the electric fence. Sheep have never been stupid.

For a hill farmer there is surely no greater joy than being at one with a dog, particularly on a day of glorious weather when all plans go like clockwork. Previous difficulties are soon forgotten, but unfortunately most sheep farms do present a dog with innumerable challenges. Soft and pampered pet sheep are a shepherd's nightmare as they do not have any respect for a collie, and will nearly always ignore a dog's efforts to move them to another field or byre. Unusually one morning after they had been let out, our two collies vanished. Though I whistled and yelled they did not return. After a time I heard muffled barking coming from one of the far sheds, and on investigation found the source of the trouble. One of the pet sheep had escaped and had been gourmandising in the feed store, helping herself to a bucket of oats. In her greed she had managed to get the bucket stuck on her head and was inanely wandering round closely kepped by the dogs. However, incarcerated in a bucket helmet she was unable to thump them as usual, and they seemed almost gleeful to be in charge for once, and were therefore not keen to leave.

While on a small and remote Hebridean island with only bikes for transport, my son Freddy and I enjoyed the companionship of two particularly cheery collies, Jock and Spring. Each morning as we were packing up the bags on the bikes ready to leave, these two game dogs leapt up from their hen watching duties or lazy slumbers at the croft's door to follow us as we biked round the island.

(opposite) Jim Hogg, with Loch Tay and Ben Lawers behind

Galloping beside us for a time, first one would vanish, and then shortly afterwards the other. Sometimes, they would reappear a few hours later, hot, panting, and filthy. Twice they rejoined us at the beach in time for our lunchtime picnic and enjoyed cooling off in the sea. After a few days we realised that while they thoroughly enjoyed our companionship, there being little sheep work at that time, they had other things on their minds, namely of a highly flirtatious nature. As we biked up to the telephone box one night with the dogs in hot pursuit, Jock took off up a track to a croft, and shortly after irate Gaelic oaths and a couple of well-aimed boots sent him yelping back with his tail between his legs. The bitch he had been pursuing had been shut away as her owner was heartily sick of the unwanted amorous advances from this love-lorn dog. At tea that evening I casually dropped it into the conversation with our hostess that it was amazing how Jock and Spring were spreading their genes round the island. 'Oh, I know,' she said with much wringing of her hands, 'it's wicked the way people can't keep their bitches under control, they are leading my poor boys astray.' Freddy and I almost choked on our scones.

Though my small pampered flock of sheep come running to a rattling bucket, I would never be without a collie. My current dog, Kim, has no need to work but would have been very good given the chance, as she is devotedly loyal and obedient. On occasions I take tourists on Land Rover trips up hill tracks to see wildlife. There are days when the fauna is unobliging and it is hard to find anything unusual. However, there is nothing like a good hill Blackie tup, or a Highland cow to stimulate the tourist's cameras and videos into action. Normally Kim stays behind. But one day she looked so devastated at the thought of being abandoned that I weakened. I was busily pointing out some far distant Blackies to a party of American tourists when I spotted a raven's nest on a nearby cliff face. Attention was quickly focused on this spectacle and our backs turned on our earlier point of interest. Ravens are fascinating birds so absorbed us for some time. I thought Kim was asleep in the back of the

Land Rover, but later when I turned round found that she had been doing her own thing and had returned with about 30 Blackies, including a very impressive tup. They stood in a neat group beautifully placed round the vehicle – perfectly poised for visitor appraisal. The collie is a dog of perceptive initiative. The Americans loved it and Kim was a temporary heroine. Once the intense photo-call was over, we fled quickly as I was rather ashamed by my dog's wayward behaviour.

The Border collie is incredibly versatile and makes a perfect companion. Ratting is a favoured pastime and they can account for many of these unpleasant farmyard pests, especially when they are poked out from under bales with a long stick directly into the path of waiting jaws. A collie does not give in easily and is usually very persistent ensuring that its line is kept up. Moonlit forays round the neighbouring district bear much unwanted fruit, and as we let our dogs out first thing in the morning we have to be watchful for black and white rapists loitering in the shadows. The doleful serenades of one local dog that comes in the small hours of the night while Kim is on heat are enough to wake the dead. Well-aimed buckets of water help to cool the ardour when the hormones are running amok, but usually make our return to sleep hard too.

Being so intelligent, collies frequently develop behaviour traits, often brought on by boredom, or close breeding. Chasing vehicles is a familiar bad habit, and nipping the backside of a suited sale rep in the yard is not a pleasant trick, but is sometimes not discouraged either. Oddly perceptive, our collies seemed to pick up our irritation as travelling salesmen always arrived at the most inopportune moments. With minds like sponges, and an intelligence that is second to none, boredom for a collie is always best avoided.

Many farm buildings bear the wounds and battle scars of teeth and claws. When shut away in a dark shed it can be very useful to carry out a little DIY joinery in order to enlarge a window on the view of the yard. Collie sculpted woodwork can be found on most farms. Kennelling varies hugely. While some dogs doss in the back of the Land Rover, others have converted barrels on their sides, to which they may be chained, and some have smart purpose-built sheds. Many just kip on the bales if they are not allowed into the house.

Collies vary hugely in their appearance and can be bare-skinned or long-haired. Some are large and rangy, while others are small and fox-like. Stealthy of movement, and with hawk-like eyes, they all work in a different way, some preferring to be up on their feet for most of the time, while others crouch low to the ground or lie eyeing up their flock hypnotically. But there has always been a regular pattern to their names: Jed, Ben, Di, Glen, Tweed, Cap, Jess, Clyde, Moss, Sly, Nell, Meg, Gyp, Fleet, Corrie, Fly, Fan, Bess and Lass. Though you may travel to a hundred farms, you will find that most collies' names are short and simple, unlike the pretentious appellations given to many other dog breeds. Long impressive names are quite out of place, and anyway would be instantly carried away on the wind of a wild Scottish hillside. This is the reason why generations of Border collies have been stuck in a rut as far as nomenclature goes. Perhaps it is rather surprising that despite their short, punchy titles, they often respond to many other words that I would be better not to repeat. This is a dog of supreme intelligence.

Dung heaps and dead things offer a great opportunity for a whole range of canine odour enhancement. It can be quite shocking what a collie will find to roll in. Then it looks terribly hurt when you are less than enthusiastic as it slinks in by the fire at night, and as the heat makes the pungent aroma ten times worse you are forced to banish it outside. Some dogs are thieves by nature, and this can be a hard habit to break. One of our collies regularly sneaked into the house to pinch the sandwiches off the side, greedily gulping them down in a second. Fed-up with this, we made some one day using the strongest curry paste we could buy and left these delicious temptations to await her. The poor dog spent the entire afternoon drinking. Obviously unimpressed by the flavours of the Orient, she did not risk thieving again.

Kim

It can be very hard to find a good all round dog that does everything required of it. Some farmers end up with many collies, each suited to a particular type of sheep work. The best dogs on a mixed farm are those that will not only single out a ewe and never lose sight of her even when she is fleeing across the heather with a huge group of others, but also help to bring in the cows. A dog that is good on the hill often has to work out of sight, so needs different skills to one working in an in-bye situation. Hill sheep are often more thrawn than their low ground counterparts and are frequently far less obliging, testing a dog's skills hugely, and the breaking away of even one ewe from a group can be enough to totally ruin the best-planned manoeuvres. The livestock farmer today has far less time to work with his sheep, and it can be hard for a dog to have enough practice.

For the past few years Jim Hogg has not only been training collies, but has also been teaching Lou Radford, an acupuncturist, both the finer, and perhaps not so fine points of working with the Border collie. 'When I first started helping Lou with her dogs, I really did not expect to see her returning, as like the dogs she has received many a swearing. But she is amazingly dedicated and has kept on coming back for more. Recently she successfully won a trial at Blair Atholl, and thoroughly deserved it.' This is praise indeed coming from Jim who does not suffer fools gladly. 'One day we had arranged to go out with the dogs and the weather was horrendous with horizontal sleet and a gale. I just assumed Lou would not turn up, but no, there she was and when I opened the door, I said, "Jesus, I would not even put a milk bottle out in this weather"', Jim laughs. We are watching him put both Lou and one of her dogs through their paces, trying to shed off two fat sheep from a large group at Mervyn Browne's farm, Milton of Ardtalnaig, on the side of Loch Tay. In this wonderful setting with the distinctive shapes of Ben More and Stob Binean at the head of the loch, and the peaks of the Ben Lawers group in front, Lou has learnt a great deal about dog training, and now comes to help Mervyn every year with the lambing.

It takes a great deal more time to shed off the two fat sheep than Lou and Jim anticipate as they test the dogs thoroughly, the flock running up and down the steep field refusing to split. Jim intervenes and brings in another dog, giving both Lou and her dog a few stray Anglo-Saxon oaths on the way. The fleeing Blackies are in uncooperative mood. Soon both Jim and the second collie are looking a trifle hot and bothered. Eventually, the flock is driven to the bottom of the field and he is left with four by the gate. He opens it to let the two large ones out, struggling to stop the Blackie lambs from following, quickly shutting the gate. With a mad dash one lamb leaps through the fence to join the others. Jim grabs the struggling creature and shoves it back into the field. Infuriatingly like a jack-in-the-box it pops back and forth several more times as if the fence was non-existent. By now the air is as blue as the sky. Not only can sheep bring out the worst in you, but they can even surprise a brilliant handler – and his dog too!

Although farming is constantly changing, some things remain the same. Where there are sheep there will always be collies, for without them sheep husbandry would be well-nigh impossible. With a great eye and style, this fantastically intelligent servant has in-built skills which are up to us to develop. The Border collie is as much a part of Scotland as its heather hills, lochs, and glens. But like us, they too have very varied characters, occasionally mirroring some of our least admirable qualities. Jim Hogg would say that we only have ourselves to blame.

You can hurl as much verbal abuse at a collie as you like, and perhaps the most amazing aspect of this wonderful dog is the fact that it will usually forgive you and come back for more. Equipped with intuition, an iron constitution, loyalty, sagacity and agility, the Border collie is surely one of Scotland's greatest products, deserving of our kindness and utmost respect.

CHAPTER NINETEEN

A Shetland Miscellany

A forklift truck towed a large Aberdeen-Angus bull in a trailer round Kirkwall harbour as we waited to board the boat from Orkney to Shetland on a fine June evening. He too was Shetland bound. Fishing boats laden with shellfish were being relieved of their cargoes as large refrigerated container lorries prepared to take them to their destination on the gourmet dining tables of Europe. A couple of drunks tottered round the pier and uttered a garbled greeting as my eyes met theirs. In front of us a group of exceedingly large tourists crammed into a very small hire car would certainly have had their pennyworth out of the weighing machine, and reminded me of four elephants in a mini.

On board P&O's ageing St.Sunniva we found the cabin surprisingly comfortable. But showering was a mistake as the boat's lurches at just the wrong moment caused the tiny ablutions department to be temporarily waterlogged, and I had to wait some time before the flood subsided and it was safe to open the door.

The remote group of approximately a hundred islands that make up Shetland lie 200 miles north of Aberdeen. First colonised in the ninth century by Norsemen, they became part of Scotland in 1469. Situated on latitude 60, Shetland is nearer to the Arctic Circle than to London. In mid-summer the sun remains above the horizon for 18 hours and 52 minutes, and there is only about one hour of half-light between dusk and daybreak. This is referred to as the *simmer dim*. Famed for its unique geology, and of international importance for many seabirds, Shetland is also renowned for its small, hardy livestock breeds that through time have evolved to thrive on the poorest grazing, amid frantic, unforgiving gale-force winds and a high rainfall.

'You will recognise our house as it is the only one with a big tree in the garden.' There are few trees in Shetland and any survivors are usually wind-sculpted, stunted, or prehistorically shaped. If the wind does not put paid to them, then the sheep almost certainly will. We knew it was Agnes and Davy Leask's croft not by the tree, but by the large variety of Shetland sheep grazing with their colourful lambs in the field in front. 'Come away in and have a cup of tea. I am in sore need of one as we've been on the go since the crack of dawn.' Agnes Leask has a very slight build and perhaps its accompanying shrew metabolism. Certainly, it is obvious that she is always on the go and not one prone to slothfulness.

Inside the warm kitchen a friendly collie bitch recuperates by the range having had a small operation. Clearly feeling better following devoted pastoral care from her doting owners, she paws at us for attention. On the wall, a large black and white photograph depicts a couple standing beside a small, neat cow with a baby sitting on its back. 'That's Fittie, who had a wonderful nature. Here in Shetland the cow was as valuable as many of the family, and could also be the provider of the roof over your head. And the baby is me,' Agnes explains.

A very ancient breed descended from domesticated cattle brought to the islands by neolithic settlers, the Shetland cow, fondly referred to as kye, like the rest of Shetland's native livestock was traditionally small, and had an impressive ability to exist on the most impoverished grazing. She kept the family in milk, reared a calf, and managed to survive great famine. Even though at the end of a terribly long hard winter she would be so weak from malnutrition that she was often unable to stand up and would have to be put in a sling until good feeding had brought her condition and strength back. Any softer bred beast would have succumbed without doubt. With a wonderful temperament these diminutive cows were of the greatest importance to Shetland families, and often ended up supplying the neighbours with milk too. 'My mother always made sure that the cow had enough to eat – without her we would have had nothing, and it was always a great tragedy if the cow died prematurely on any croft. When our cow had calved, we used to rip up a piece of hessian sacking and tie on a cover with coir string to keep her warm. Sometimes we even lit a fire in the byre. You could hardly see the beast for all the peat smoke, but it was important to look after her. To keep her back end clean, we used to trim the hair off her tail and laid it on the wall top. It was an old belief that if you threw that away, you also threw away the luck,' Agnes explains.

Today, the Leasks no longer have a cow but still keep a flock of traditional Shetland ewes. 'I have always been keen on sheep, but it is the really old-fashioned ones that we keep here. They are clever and hardy, and I love all their different colours.' Katmogit, yuglet, bersugget, smirslet, bleset, brandet, krunet, snaelit, flecket and kranset – these tantalisingly descriptive words are just a few of the names used to describe the different colours and patterns found in Shetland sheep. Like the landscape, the sheep reflect the earthy colours of bracken, rocks, lichens, and the constantly changing moods of Shetland's vast skies.

Agnes whistles at the collie and sends her out to gather them in. Soon they are puffing up the hill with the dog at their heel. A patchwork of muted colours, they stand and stamp in defiance. In amongst them there is a four-horned ram. 'It is my belief that this is a throwback to ancient ancestors, and I think if you look far enough back you will find that many of the sheep would have had four horns. Not everyone would agree with me,' Agnes explains. Shetland sheep are divided into two different categories, the Flock Book Shetlands, and the more primitive variety favoured by the Leasks which remain unregistered but are very representational of the old type of hill sheep once found all over Shetland. Many of the older type are horned.

'When I was young you only clipped something that had very matted wool or maggots, although we don't have a problem with the latter really. All our sheep were *rooed*, or plucked, using a forefinger and thumb. It gave a far better finish than the electric clippers which leave a sheep far too bare.' The Shetland sheep produces very high-quality wool. In Agnes' youth Shetland's thriving knitwear industry was a source of income for most of the island's crofters. 'Instead of doing my homework after school,

Katmogit Shetland sheep, Skeldaness

I would be sitting knitting patterned children's mittens. I could do a pair in an evening and I would be paid about 4 shillings and 6d for them. We used to sell them direct to the men in the army camps, and they probably sent them home as presents for their wee ones. Even after the war there was a huge demand for Shetland knitwear, and we all had patterns stored in our heads for many lovely designs. But then the knitting machine arrived on the scene and it was really the death knell to the hand-knitting industry,' Agnes explains.

'My mother bought some Shetland hill lambs from an old lady who had walked them all the way to Bixter Market from a distant croft, trailing her long skirts as she went. I desperately wanted my own sheep and asked if I might have one. She agreed and I chose a black one with a star on her face, and that was the beginning of a life-long passion. I had that ewe for 10 years, and when she was taken away I was paid £3 for the skin which was a fortune then. Shetlands are so hardy, and they can go on breeding for years. You won't find a ewe that is a better mother,' Agnes says. Davy laughs, 'When Agnes went away to a meeting she told me to make sure that one of our old ewes did not go in with the tup. However, she just leapt out and wandered off down the road and jumped in with the neighbour's tup instead. We reckon she is nearly 20 years old, and she still has a lamb. When she sees a strange dog coming anywhere near, she will jump out of the field and does not return until it has gone. They say sheep are stupid!' They both laugh. 'I really love animals and I have always been bothered by all this transportation. Some livestock travels for miles to the other end of the country, and often abroad too. I don't mind when sheep are slaughtered locally or on the croft, and I really feel that there should be more government support for small local abattoirs. We would benefit from that, but I can't see it happening.' Agnes sighs.

In 1958 the Leasks beggared themselves to buy their croft at Cott in Weisdale. It nestles under the hill overlooking the Weisdale Voe, and today is a far cry from the tumbledown property they first acquired. 'You took on a croft, you had a child, you were penniless. It was a question of give up and admit defeat, or carry on. We carried on. Neither of us regret it,' Agnes tells me. She has been Vice President, and President of the Crofter's Union for many years, and in this role has had to take on some feisty politicians. Though she is very small and lightly built, Agnes certainly has a big personality, a great mixture of humour and kindness, but there is no doubt that she is fiercely determined when necessary. 'I really enjoy flying off to meetings in Inverness and further south, but I have had to cut through a large amount of red tape. That is one thing I simply can't be doing with. Most of it is just stupid policies dreamed up by politicians that have absolutely no bearing on real life. Some of these new regulations have to be fought or else none of us would survive out here, so I don't hesitate to fight the crofter's corner if I have to.' She strokes the collie, one of eight dogs that are clearly another passion. 'Agnes' grandfather twice fought the laird and won,' her husband Davy adds, 'so I know where she gets it from.' They both laugh. 'Crofting is important here, and I feel strongly that it should still be possible to continue in a small way. It is important too to keep these old type of sheep going, that is one of the many reasons why we still have them. Mind you, I wouldn't be without them, and I feel we must keep on breeding all these lovely colours.' I noted that it is not only Shetland's livestock that is small and hardy.

Tammy Fraser, aged 95, had just lost his wife. They had been married for over 68 years. Despite his recent bereavement he agreed to a meeting, and I was welcomed at his flower-covered doorway like a long lost friend. Tammy Fraser is the kind of man who is very hard to leave, and I regretted that I only had a few hours to spend in his company. His smile would brighten the most inclement of Shetland days, and his giggle was so infectious that he had me totally under his charming spell. Mention Shetland cattle to anyone in the know and his name instantly crops up. 'When I was a boy all I thought of was chooks, kye, horses, and sheep. But eventually it was Shetland kye that were my thing, and I had them all my life until age forced me to give them up. In 1937 I took on the Glebe here at Garderhouse,

and I put four Shetland kye on to the ground. They were pure bred: grey and white, black and white, and pure black in colour. They were the foundation stock for my Glebe Herd. Then the croft was very small, but we kept on reclaiming a little more land and I added kye as we did so. They were wrongly thought of as a dairy cow. However, this was not the case as they are a dual-purpose breed. They could rear a beef calf, as well as keep a croft supplied with milk even off the poorest of ground. They were the only breed that could do this. It was very sad to observe them lose popularity and see only the black and white ones sell well at the market in Aberdeen. I began to concentrate on breeding this colour, and soon had fixed a type. Now the other colours including red, and red and white, are being bred again.

'My kye were lovely animals, and I was always very keen on showing, and had my name on the cup at Walls Show 21 times.' Tammy's eyes light up. 'Oh, they were such lovely kye! I remember one bull calf, Glebe Rasmie, he really filled my eye altogether. They sorely needed a bull up in Fair Isle, but I said that I could not part with him. But they told me that if I didn't there would be no calves up there the following year. So away he went to Fair Isle for two years before going on to the Department of Agriculture at Inverness. I never did get him back.'

The Shetland cow produces such an excellent quality calf when crossed with other larger breeds that this was nearly her downfall, as for a time fewer pure calves were being bred. As popularity for the island's cattle dwindled, Tammy realised that something had to be done, and with the help of the local doctor's son, a Shetland breed enthusiast, Hugh Bowie, and Robert Ramsay from South Collafirth, he re-established the Shetland Cattle Herd Book which had been abandoned for many years. Eventually the Department of Agriculture who were working on improving livestock for the Highlands and Islands became aware of the possibilities of the Shetland cow and took some cattle from the Glebe as

Agnes' sheep, Davy's lorry

Shetland cow and Shetland ducks, Burland Croft Trail, Trondra

well as a few others from the island to start the important Knocknagael Herd. Based in Inverness, Jimmy Dene's work there with them proved to be of vital importance, and together with the Rare Breeds Survival Trust, and Tammy Fraser, helped to save them from dying out altogether. In 1973 when the Trust was first established, Shetland cattle were placed on their priority list. In 1982 there were only about 100 registered females left. Today there are once again a few flourishing herds on Shetland, in particular those of Robert Ramsay and Evelyn and Norman Leask, as well as several on the mainland, including a large herd belonging to Mr and Mrs J. P. McCaig, staunch supporters of the breed from Lanarkshire. Today most Shetland cows are much larger and better nourished than their predecessors, simply because they have access to better feed and no longer exist on a system of controlled malnutrition.

'Today it's the same coo, but with richer feeding there's more growth than in the past, but that's surely a good thing. Most of them are now on good keep all the time, yet still retain their Shetland characteristics. We have tried to keep them pure, but have been accused of bringing in Friesian blood you know, but it's not true.'

On the small island of Trondra, Tommy and Mary Isbister keep examples of all Shetland's native breeds. Two more dedicated people would be hard to find. Their pied cattle looked particularly eye-catching as they grazed a daisy covered pasture, while in the background a large flock of Shetland geese dabbled on the sparkling sea. Shetland's hardy native goose is sex-linked, and unlike many other birds its gender can be recognised as soon as it is hatched. Their cows, which old Tammy Fraser had described as 'some of the best examples on the island', were proof of the results of good quality feeding and excellent husbandry. Sleek, healthy, and fat from summer grass, the modern Shetland cow is certainly not malnourished. 'It was in 1979 that we tried to find ourselves a Shetland cow because I had grown up with them. I remember ringing old Tammy, but he said we were too late. It was not until 1981 that we finally found one and were thoroughly vetted by her owner to see that we were suitable people to keep her', Tommy tells us.

Tommy and Mary have opened their 60 acre crofts to the public. Mary has a lovely warm manner, and clearly loves to share the pleasure they derive from their crofts with the visitors. They have a few beautiful Shetland ponies and keep a couple of pet lambs every year especially for visiting children. Their land is farmed traditionally, growing small areas of old-fashioned crops such as bere barley, a primitive grain that has been grown in the islands since the Iron Age. Black Shetland, Foula Red, and Blue Heart sound like unpleasant diseases, but are in fact delicious old varieties of potatoes that were once prolific here and are now scarcely seen. A blue-coloured cabbage is reputed to have arrived on Shetland with Cromwell's soldiers. The Isbisters sell a little seed from their vegetables as they feel that it is important not to lose them.

Their crofts slope down to the sea and are vulnerable to the westerly gales that frequently drive in with great velocity plastering everything in salt. However, the azure sea was calmly hiding its fickle nature as the sun beat down on their flock of primitive Shetland sheep. Like the Leasks, the Isbisters keep the old type of sheep, and have some beautifully coloured animals that originated on Shetland's most westerly island, Foula. Due to its isolated location the sheep there have remained remarkably unaltered over the years.

An up-turned, un-seaworthy boat had a new role as the roof for a stone-built poultry house. The contented occupants of this most attractive dwelling scratched around in the yellow flag iris or dust bathed on the path. With great dedication the Isbisters have revived Shetland's original breeds of poultry. Almost every croft in Shetland would have had a selection of fowls to provide eggs and meat. The island's earliest hens were thought to have consisted of two distinct types: a larger which has a characteristic tuft or 'tapp' of feathers on its head, while the smaller has a very tiny, neat comb. Both types are very varied in colour. Mary has thoroughly researched their history, and today continues to breed both varieties, always trying to bring out their best characteristics. While on holiday on Foula

they discovered the last of a line of Shetland ducks and persuaded their owner to part with her. When they found a few more on the east of Shetland they started a breeding programme, culling out the poorer drakes until they had ducks of a type that must be very similar to the old ones. Though genetically the scientifically-minded would perhaps dispute their theories, there is no doubt that their efforts have resulted in the resurrection of the Shetland hen, duck, and goose. All have almost identical characteristics to their early Scandinavian predecessors, having been naturally raised through the Isbister's loyal efforts. Shetland poultry is much in demand, for some of their birds have been sent to the South of England, and a few have even ended up in America.

While the Isbisters work exceedingly hard farming their crofts in a traditional and environmentally friendly manner, and are largely self-sufficient, other income has to be found to supplement their keep. Tommy is a true craftsman and makes beautifully designed fiddles, and traditional Shetland boats of the old, Norse type. 'I always feel that the next one will be better. An old man once said to me, "You'll never be happy with them", and I suppose he is right. I always see something that I want to change,' Tommy explains. 'He often rushes his food so that he can get on and make his fiddles in the evenings. He is brilliant at this, but he can't make the tea,' Mary laughs. There are about two hundred hours work in a fiddle, and so it is really a labour of love that is very dependent on how the spirit moves him. Tommy passes me one of the fiddles, which is so skilfully made that I can see this is a man with a great gift.

It was a grey and dismal evening as we took to the west to visit the Ridlands, at Bridge of Walls. The white forms of whooper swans and their five cygnets shining out of the sad mist on a large roadside loch were a brief highlight. The whooper swan rarely breeds in Britain. The Ridlands have been breeding Flock Book Shetland sheep for many years now, and Brian and Fiona have had considerable success with their tups. The Shetland, a member of the northern short-tailed group of primitive sheep, is one of the smaller breeds, and was traditionally bred for the hill. It has always been renowned for its hardiness and fine quality wool that is on a par with that of the Merino. The Shetland Flock Book Society was formed in 1926 in order to monitor the constant improvement of the wool, and to encourage a hardy, healthy, and strong flock with a greater mutton yield. The Shetland produces delicious meat and an excellent commercial lamb when crossed with many other breeds. She is frequently crossed with the North Country Cheviot, and the female progeny from this cross can then be put to a larger terminal sire. Today, the majority of Shetland Flock Book sheep are white, although moorits, blacks and katmogits (dark under-parts with light upper-parts) are also quite common, and other colours may be registered in the Flock Book provided they are of a high standard. More unusual coloured animals which need to be preserved, and many of those on the mainland were, until recently, listed by the Rare Breeds Survival Trust. However, with over 5000 sheep now registered in the Flock Book, the Shetland has been removed from the Rare Breed Survival Trust's list, although they will continue to monitor their status as a fledgling 'former rare breed'.

From the age of two, Brian Ridland was out with his father Jack working with the sheep. 'I have always been passionate about them, and when I was a youth used to go off with Tammy Fraser round farms and crofts to see other people's sheep. You can learn such a great deal from the older ones, and

Tammy has that jolly way about him that a kid can relate to. When he was a boy he used to work with my great-grandfather. I was 11 when I acquired my first tup. Much of the stock on the farm today will be related to him. We concentrate mainly on white sheep. Currently there is a greater demand for white wool, although the wonderful range of natural colours available from the Shetland, are popular with knitters and spinners all over the world. Americans particularly love coloured wool. Here on the island we don't deal with the Wool Board, and our clip is sold directly through Shetland Wool Brokers. There has always been a market for it and I try to concentrate on keeping the quality as it is important to maintain a high standard.'

Flock Book Shetland ewes are mostly polled, but the tups have well-rounded horns that must not grow too close to the face. Sometimes they have a slight black stripe through their horns which can be very attractive. 'For years I have tried to breed good even horns, and a strong character in the face with alertness of eye. However, once you get one end right then you discover that there is perhaps something at the other end that you don't like. Occasionally the sheep can also have a slight black speckling on their face or legs, but you certainly don't want any black in the fleece of a white animal. Longer, coarser wool on the breech is acceptable so long as the overall appearance of the fleece is uniform and shows fineness. Hardiness is vital, and I do not like to breed from anything that won't withstand the weather. The animals must be strong with good breadth throughout, and defined height at the shoulder. I have worked very hard on our sheep and now feel that we have achieved just about what we are looking for,' Brian explains.

He won his first show in 1983 and has been winning ever since. Tups that are chosen earlier in the year for the show ring are rooed in the traditional manner. For the show ring Brian agrees with Agnes Leask that clipping leaves the animals too bare, although all the other sheep on the farm are clipped with the machine. He was busily rooing a very handsome tup, and demonstrated the method of neatly breaking the wool with a finger and thumb at the point where the new wool is growing in from underneath. Sometimes it can take up to eight hours to complete. 'He is often out in that shed,' laughs Fiona, 'Sometimes he and his father are in there for hours and hours in the evenings, working away and discussing sheep. I often wonder if he will ever come back into the house. They are infatuated with their sheep.'

Surviving purely off the farm's income is impossible in today's current agricultural climate. Brian has another part-time job working for the Shetland Council as maintenance supervisor in their infrastructure department, and oversees all council plant and vehicles. He freely admits that he could not survive on the farm without this, and has to work hard in order to be able to enjoy the sheep. Luckily, his job is fairly flexible and he managed to stay at home the following morning to take us out to Silwick where they keep 20 katmogit ewes belonging to Fiona. Following behind his pick-up we were much amused by the frenetic activity of his collie, Cap, in the back. Full of excitement, he was flying round and round in circles like a record on a turntable, occasionally reversing direction.

Shetland ponies, Unst

The dismal mist had cleared away leaving a wild and squally morning. The sun was having a passionate dispute with the racing clouds fleetingly appearing to brighten the landscape, only to be quickly put back in its place. The Ridland's katmogit flock are kept separately from the rest of their sheep on a rugged headland at Skeldaness where they survive on no extra feeding, even during the winter, only supplementing their diet with seaweed off the shore. On a wave-swept headland before us, a large party of cormorants dried their wings in the wind while others nested on the knife-edge of a vast sea-stack that rears up out of the Atlantic swell. This westerly side of the coast is a fascinating mixture of pink and red cliffs, vertiginous coves, wind-sculpted stacks and natural archways, carved out by a combination of tide, time, and weather. We thought that Cap would be far too dizzy to round up the sheep, but as soon as he was bidden, he was off towards them and shortly had the ewes and lambs hurtling in our direction.

Katmogit Shetland sheep are perhaps one of the most beautiful of all with their dark underbellies, and tan colouring round their faces and legs. Brian and Cap drove them neatly on to a small rocky promontory jutting out into the angry sea. With a square topped stack behind them and large blotchy clouds sending bands of rain across a light-streaked vista, the view was almost ethereal. Pictorially, this was one of the best moments of our journey. Like the North Ronaldsay, these sheep are perfectly suited to their rugged landscape.

There are records of ponies in Shetland for over 2500 years, and during that time they have been the mainstay of the island communities, providing vital transport and assistance in many tasks on the crofts. Laden with panniers of fish, peat, and seaweed, these clever animals proved invaluable. The Shetland pony is the smallest of all the native breeds, and due to the isolation of its homeland has remained remarkably pure. However, their diminutive size was also to have a very detrimental effect on the breed, for when the Mines Act was passed in 1842 banning women and children from working down them, the demand for ponies as draught animals became immense. Many left the wild and rugged islands for a miserable new life spent in almost total darkness, deep down the mines, where their incredible strength was used to haul great loads of coal up the dank, unhealthy shafts. Many were killed or severely injured in the frequent accidents that occurred when pits collapsed, and were hastily replaced with others. At that time it was the smallest males that were in the most demand as mares were often far less reliable during their breeding cycles. The Marquis of Londonderry, a mine owner, realised that removing so many good ponies from Shetland was going to cause a deficit and loss of quality throughout the island's dwindling herds. He set up a stud on the island of Bressay using some of the best remaining stallions, and today many ponies still have the influence of his good work through their pedigrees.

After driving to the north of mainland Shetland, there are two car ferries to travel on before reaching Unst, Britain's most northerly-inhabited island. This remote and unique 12-mile-long island has been synonymous with the Shetland pony for countless generations, and still retains some excellent studs. Incredibly varied, the landscape of Unst supports a rich flora and fauna as well as its impressive herds of ponies and flocks of native sheep.

Brian and Margaret Hunter and their son Peter have been breeding ponies here for most of their lives. Brian worked with the Crofter's Commission for many years and was the commissioner for Shetland, as well as being Honorary Chairman of Shetland Marts. Margaret has recently written a book about Shetland ponies. 'At the moment the Shetland is very popular, but it would not matter if they weren't as we would always find reasons for keeping them,' Brian laughs. The Clivocast Stud ponies are all bred in natural surroundings with only a little supplementary silage fed to them during the winter. 'My father was brought up on the island of Fetlar, and they needed a great many ponies there for carrying all the peats. They were of vital importance for peat flittings all over Shetland. Then

Brian Ridland's Shetland tup

a pony under 34 inches in height was considered small, but now many miniature Shetlands are regularly registered under that height, and there is a considerable demand for them for showing and as children's first ridden ponies,' Brian explains. The standard ponies are between 34–42 inches in height. 'We are fortunate that there is a good market for ponies. Many people come to see them, and there are Shetlands all over the world which have adapted amazingly well to totally different climates and surroundings. I have judged them at the Sydney Royal Show in Australia, and it was interesting to see that the ponies there were pretty similar to our own, although in the heat the black ones tend to fade to a rather strange brown. I love all the flak involved with judging, and remember once judging in Northern Ireland. There was a woman there with a pony that was reputed to have cost £5000, but it did not show itself well as it was bored, tired, and fed-up, so she didn't win, and she was not at all happy about it,' Brian laughs.

The Shetland pony may be any colour except spotted, though black remains a popular colour in the show ring. 'It is much easier to judge the pure black ones as when an animal is piebald or skewbald, or perhaps has one whiter leg than another, it can be very deceptive as the white will always draw the eye.' During the winter months, Shetland ponies grow a double coat with long guard hairs that help the water to run off. Manes and tails are always kept long and thick adding further protection from the elements. The Hunter's ponies were scattered over a wide area of Unst. Most were in the throes of moulting out their thick coats and stood grooming one another's necks, revealing a smooth silky coat underneath. Despite being on a heath close to the road, none of their ponies showed any signs of wanting to nip. Titbits have been the downfall of many a good small pony and can spoil an otherwise excellent temperament. In relation to its size, the Shetland is the strongest equine there is, and so firm handling and basic training are important. Incredibly intelligent and quick to learn, good habits are easily established, and it is therefore a shame that occasionally they have earned a reputation for being devilish. In the right hands they are wonderful animals. Each year when the Hunters wean their foals in October, they bring them inside for a short time in order to get them used to being handled.

'For me a pony must have a pretty head and good bone. A smart, straight action is also vital. They should have a lively trot and good round feet. Here, we do not have a problem with ponies developing laminitis, unlike on the mainland where much of the grazing is far too rich for them,' Brian tells me. As a group of coloured ponies trotted out towards us sending up a spray from the boggy ground, their cheery disposition was another very obvious attribute. In recent years Shetland ponies have become immensely popular. With their incredible stamina they make supreme driving ponies. For many years The International Horse Show at Olympia has staged a Shetland Pony Grand National. Ponies have to qualify for this event at various other shows round the country. It is always full of hilarity and fun, and without doubt proves the versatility and courage of these diminutive equines. If ever a pony had a sense of humour, then it is the Shetland.

There are thriving Shetland pony studs all over Scotland, while presently there are approximately 1000 registered ponies on the islands of their birth. This phenomenal pony is loved and appreciated by a world-wide audience. We passed many herds in a large variety of landscapes. Like the sheep, the ponies occasionally add seaweed to their diet, and we frequently saw them paddling in the sea to cool off.

On Unst the Hunter's mares and foals were divided into groups, each with a different stallion. We found a good place to cook supper over our primus stove, scrambling the rich eggs Margaret had kindly given us. As we were preparing the food, we watched their ponies grazing quietly round the remains of abandoned stone crofts. Unst's population has now fallen to 800, but at its peak was about 3000. The ruins are testimony to its depopulation. The smell of cooking brought a line of hurrying domestic ducks from nowhere. They gladly relieved us of the remains of lunch before they moved off to guddle loudly in the roadside ditch. It was midsummer's night and all the B&B accommodation on the island was full. Margaret had booked us into Britain's most northerly Youth Hostel.

As we headed back there we stopped beside a small lochan surrounded by moorland, enthralled by the intense activity in so small an area. A pair of red-throated divers fed their chick, carefully guarding their treasure from the attentions of a pair of Arctic skuas. More refined and streamlined than the great skua, their distinctively pointed tail and lighter elegant shape was clearly apparent. Whimbrel battled with the skuas too, driving them away from their young hidden deep in the heather and cotton grasses, while calling in warning. Suddenly we caught a glimpse of a small starling-sized bird flying across the water, and through binoculars saw that it was a female red-necked phalarope. Unlike other birds, it is the female of the species that has the brilliant plumage, and unusually they are surprisingly tame and therefore easily approachable. We quietly waded out into the water until we were close to this perfect little wader, watching as it plucked tiny insects from round the protruding rocks. Behind us a medley of snipe reverberations and the plaintive calls of golden plover carried over the moor, to which curlew added their beautiful, melancholy music. Soft drizzle fizzed the water as beads of moisture added seed pearls to the immaculate red and grey plumage of the phalarope.

Dawn breaks clear and windy. Situated at the very top of Scotland, Hermaness, the most northerly bird reserve, is home to one of the largest colonies of great skuas in the world. More usually referred to as the bonxie, during the breeding season these pirates of the air are surprisingly savage in defence of their territory. The path at Hermaness leads the visitor close to the middle of the bonxie's maternity wing, and it is hard to proceed unnoticed. Carrying a stick above the head stops the birds from hitting their target, and also saves a headache. Large and imposing, they swoop on us with great accuracy until we are well out of their breeding zone. In 1831 there were only three pairs of bonxies left at Hermaness due to the activities of egg collectors and taxidermists. Today, the reserve has at least 650 pairs, and running the gauntlet of their aerial terrorism makes even the slowest walker put on an impressive spurt.

The wild tract of open ground that leads up to the vast sea-girt cliffs of Hermaness is a wader's paradise. In June, the glorious blanket bogs and heather-clad moorland teem with life. Once the hazards of the bonxies have been negotiated the naturalist can sit back to take it all in. Amid a tapestry of orchids and sundew, coloured moor grasses and sedges, the sounds and sights of breeding birds awaken every sense. Dunlins trill over the bogs, while high above snipe display tirelessly. Almost sad, the intermittent voices of the bespeckled black and golden-yellow plovers calling from a sea of creamy floating cotton grass, beckon insistently.

As we come over the edge of the moor, the sea slides into view. A giant stack iced with white gannets emerges, while the surrounding sky and sea boil with their activity as they cruise amid gliding fulmars and the clamour of over 20,000 guillemots. Guarded by Scotland's most northerly lighthouse, lonely Muckle Flugga, the massive cliffs seethe with avian jostling. The sheer grassy slopes dropping straight into the sea are dotted with puffins. They emerge from their burrows crooning and fussing over black fluffed chicks inside, before launching into the wind in search of sustenance. Sheep and their lambs graze the short-cropped turf or lie cudding in the sea-campion.

The bonxies are busy. Cashing in on other's fishing trips, they swoop down on the returning mariners forcing them to abandon their catches in a panic. As we sit on the headland on the crest of a great bastion of rock, the aroma of guano hangs thickly in the atmosphere, and away to the north a dense bank of dark cloud, ink on blotting paper, rolls over the water. Framed by deep blue from the west, a ewe and lamb appear over the brow of the hill. Back-lit by silvery light, she stretches while the lamb suckles, wind teasing their wool. Midsummer, on Scotland's edge, embraced by birds, we have reached the end of this journey.

(opposite) Hermaness with Muckle Flugga Lighthouse

Glossary

Bogie roll – type of tobacco
Bothy – traditional stone cottage
Brockie – black and white facial markings of sheep
Buchts – gathering pens for sheep
Cairn – heap of stones on top of a hill
Cleit – small dry stone structure used for storing food – St Kilda
Coir matting – strong fibre made from coconut
Coup – to tip up
Crabbit – bad-tempered
Cruisie lamp – paraffin lamp used for lighting
Croft – small, usually tenanted farm in Scottish Highlands and Islands/ a small holding
Cromack – traditional shepherd's crook made from tup's horn
Disbudding – removal of horns
Dominie – schoolmaster
Drumming *(of snipe)* – vibrating sound made by male snipe's stiffened tail feathers during its breeding display
Ewe – female sheep
Fank – sheep handling pen
Garron – Highland pony used for carrying deer off the hills
Gimmer – two-year old female sheep
Guddling – catching fish with the hands, or turning over stones in search of small fish
Hay-hake – rack for feeding hay in field
Heft – sheep that have become attached to a particular area of hill ground
Hey you Jimmy bunnets – novelty tartan hats with red hair attached to them
Hog – young sheep between being weaned and shorn for the first time
Hope – a hill
In bye ground – land fenced between fields and open hill
Knowe – hillock
Laminitis – inflammation of the hoof – a serious complaint in over-fat ponies
Lazy-bed – patch of cultivated ground used to grow potatos or other vegetables
Machair – rare, low-lying, lime-rich, sandy habitat found on north-western seaboard of Scotland and Ireland
Mastitis – inflammation of the udder
Moll – girlfriend
Moorit – reddish brown colouring
Mouflon – wild sheep of the mountains of Corsica
Piece – sandwiches or other food taken to eat outside
Plantie cruz – circular area of dyking for growing vegetables and seed (North Ronaldsay)
Polled – without horns
Pund – term used on island of North Ronaldsay for sheep fank
Raddle – sheep marker
Rigg – a ridge of high ground
Roo – to pluck sheep's wool
Shank – downward spur of a hill
Snitcher – noose used on nose of bull to aid control
Stane-dyke – wall made of stones
Steading – farm building
Stell – circular dry stone enclosure for sheep
Tangle – seaweed
Tatties – potatoes
Terminal sire – tup or bull used to produce meat carcass
Timmer-tuned – unmelodious, unmusical
To greet – to cry (Scots)
Tup – ram, uncastrated, male sheep
Voe – Shetland term given to seawater inlet similar to Scandinavian Fjord
Wether – castrated male sheep